Intellectual Assets for Engineers and Scientists

Intellectual Assets for Engineers and Scientists

Creation and Management

Uday S. Racherla

CRC Press is an imprint of the
Taylor & Francis Group, an **informa** business

CRC Press
Taylor & Francis Group
6000 Broken Sound Parkway NW, Suite 300
Boca Raton, FL 33487-2742

Printed on acid-free paper

International Standard Book Number-13: 978-1-1383-2065-9 (Hardback)
International Standard Book Number-13: 978-1-4987-8847-2 (Paperback)
International Standard Book Number-13: 978-0-4294-3691-8 (eBook)

Library of Congress Cataloging-in-Publication Data

Names: Racherla, Uday S., 1954- author.
Title: Intellectual Assets for Engineers and Scientists: Creation and Management / Uday S. Racherla.
Description: First edition. | Boca Raton, FL: CRC Press/Taylor & Francis Group, 2019. | Includes bibliographical references and index.
Identifiers: LCCN 2018025075| ISBN 9781498788472 (pbk.: acid-free paper) | ISBN 9781138320659 (hardback: acid-free paper) | ISBN 9780429436918 (ebook)
Subjects: LCSH: Intellectual capital. | Patents. | Copyright.
Classification: LCC HD53 .R334 2019 | DDC 658.4/038--dc23
LC record available at https://lccn.loc.gov/2018025075

Visit the Taylor & Francis Web site at
http://www.taylorandfrancis.com

and the CRC Press Web site at
http://www.crcpress.com

To my Spiritual Teacher

Who taught me

"Love All, Serve All"

Contents

Section II The Full Landscape of Intellectual Property Rights

Section III Strategic Intellectual Property Management

List of Figures

List of Tables

Preface

Intellectual Assets (IAs) protect creative works, inventions and innovations on a legal basis, and create *tangible value* to customers, companies, employees, investors, and economy. Devoid of IAs, the inventors, innovators, entrepreneurs, and companies would find it extremely difficult to safeguard/recover sunk costs, let alone gain the full economic benefit of their creative investments. Thus, without appropriate IP protection, an individual/firm that invested considerable time, effort and money in developing *a new product/service*, would be at a disadvantage and risk compared to another individual/firm that just markets an *imitation product* at minimal cost, avoiding up-front investment, development time and incentive compensations to its employees. Therefore, scientists and engineers engaged in creative works, inventions and innovations must be aware of the full range of *Intellectual Assets* (IAs) and *Intellectual Property Rights* (IPRs) and understand—as part of the free market enterprise system—how to *use them to create strategic competitive advantage, wealth and value.*

Unfortunately, not many scientists and engineers—interested in creating start-ups or working in established companies—possess a clear understanding of the *full range of* IPRs *available to them,* let alone be able to use them in a *strategic manner.* As a result, we often find startups committing precious capital and rapidly executing business ideas without due-diligence on IPRs. Needless to say, such fledgling businesses expose themselves to high risks such as *infringement* of someone else's work or succumbing to imitations in a short time. The risks are even greater for larger companies. Indeed, when large companies lack IPR-savvy workforce, they face many risks such as: launch delays, product recalls, manufacturing shut-downs, infringement risks, punitive damages, erosion of profitability, loss of market share, loss of strategic IPR licensing/sale opportunities, low intellectual capital valuation and market capitalization. Therefore, this author believes that scientists and engineers must know that their job is not done just by making discoveries or inventions and participating in innovation commercialization. *They also need to be IPR-savvy and value-management centered, to partner with business to achieve sustainable competitive advantage and high valuation of their firms.* These are the urgent needs and demands of the twenty-first century due to globalization and knowledge economy.

Having led and managed global R&D programs in major Fortune 500 companies for many years, and being involved in inventing, patenting, innovating and commercializing technologies around the world, this author recognized that a major contributing factor for this problem is the lack of an easily understandable, business-relevant, comprehensive book on *Intellectual Property Rights* (IPRs) for scientists and engineers. Indeed, there are many outstanding books on IPRs dealing with complex legal matters written for students, professors and researchers of law and economics. However, *this book is unique in that it is authored by a senior business executive, technological innovator and management researcher—to provide a unique, easy-to-understand, value-management perspective to scientists and engineers.*

It has long been recognized by governments and organizations that *innovation* is pivotal to achieving *competitive advantage* and *sustainable growth.* Further, globalization and the knowledge economy have been continually fuelling the need for *innovation* of reliable, convenient, effective and affordable products or services to meet the growing needs, wants, challenges and opportunities in the world. Not surprisingly, therefore, we find many individuals, companies and nations today relentlessly working on *innovation* and *seeking* IPRs, spending valuable time and resources.

According to the ESA-USPTO study, the importance of *Intellectual Property Rights* (IPRs) in the US economy is shown to be as follows:

> *Innovation* protected by IP rights is the key to creating new jobs and growing exports. *Innovation* has a positive pervasive effect on the entire economy, and its benefits flow both upstream and downstream to every sector of the U.S. economy. *Intellectual property* (IP) is not just the final product of workers and companies—every job in some way, produces, supplies, consumes, or relies on *innovation, creativity,* and *commercial distinctiveness.* Protecting our ideas and IP promotes innovative, open, and competitive markets, and helps ensure that the U.S. private sector remains America's *innovation engine.*
>
> *https://www.uspto.gov/sites/default/files/news/publications/IP_Report_March_2012.pdf*

This recent study reported that IP-intensive industries *directly/indirectly accounted for* 27.7% of all the jobs (i.e., about 40,000,000 jobs) in the U.S. Economy in 2010. Further, this study claimed, "Average weekly wages for IP-intensive industries were $1,156 in 2010 or 42% higher than the $815 average weekly wages in other (non-IP-intensive) private industries." Finally, this study also asserted that IP-intensive industries accounted for 34.8% of the U.S. Gross Domestic Product (GDP), *in 2010.*

Accordingly, this book addresses three fundamental questions on IPRs:

Section I: What are the *underpinnings and fundamentals* of Intellectual Property Rights (IPRs) that every innovator or entrepreneur ought to know? (Chapters 1 and 2).

Section II: What is *the full landscape of* IPRs available for the protection of different types of Intellectual Property (IP) in various countries? (Chapters 3 through 7).

Section III: What are the *different strategic approaches* on IPRs for value creation, capture, and extraction? (Chapters 8 and 9).

In Section I, Chapter 1 deals with the *underpinnings of* IPRs such as *Intellectual Capital* (IC) and its relationship to IPRs. It also elaborates the importance of IPRs to companies and nations in terms of Wages, Employment and GDP based on the ESA-USPTO Study and OHIM-EPO Study.

Chapter 2 introduces the reader to *fundamentals of* IPRs such as the *different types of* IPR Systems in USA, Europe, China and India, and historical differences in the IPR systems. In addition, it outlines International IP Treaties/ Agreements on *Patents, Copyrights* and *Designs.*

In Section II, Chapter 3 deals exhaustively with all key aspects of *Patents.* Similarly, Chapter 4 treats the subject of *Copyrights* in a detailed manner. Chapter 5 describes *Trademarks* in all its important aspects. Chapter 6 elaborates other forms of IPRs such as *Industrial Designs, Plant Varieties, Geographical Indications, Semiconductor Integrated Circuits Layout-Designs,* and *Traditional Knowledge.* Chapter 7 deals with *Trade Secrets,* starting with its definition under the *Uniform Trade Secret* Act (UTSA), Differences between IPRs and *Trade Secret* Protection, Rationale for *Trade Secrets,* Examples of a *Trade Secret,* Legal Basis for Enforcement of *Trade Secre*t Protection, Remedies in Cases of *Trade Secret* Violations.

In Section III, Chapter 8 teaches the *Intellectual Property (IP) Management Strategies and Tactics.* In particular, it describes the IP Strategies used by firms for sustainable competitive advantage and growth—defensive, cost control, profit center, integrated, and visionary, as well as defensive and offensive patent Strategies.

Finally, Chapter 9 deals with the subject of IP *valuation* and describes its purpose, key strategies for successful intellectual asset (IA) management, and common methodologies used for IP *valuation,* in addition to pointing out the issues and challenges in IP *valuation.*

Acknowledgments

This book is an outcome of the diverse experiences of an individual who spent a highly productive career in Fortune 500 companies, and then transitioned into academia. At the outset, I thank my alma mater, the Indian Institute of Technology Kanpur (IIT-Kanpur) and the Industrial Engineering and Management Department, that gave me an exceptional opportunity to develop and teach new courses and conduct original research on *innovation*, IPRs, *technology management*, *strategy*, and *sustainability management*.

This book resulted from a graduate level course on "The Intellectual Property Rights Management for Value Creation and Value Capture," that I created and taught for five years at IIT-Kanpur. This course became so popular that the student enrollment increased from just 11 students the first year, to more than 100 students every year after that. I thank each and every STEM student who took my course and participated in lively discussions and insightful case studies. Indeed, they are my inspiration to write this book.

Next, I thank my students—Mukul Joshi, Sarabjoth, and Shivam Gupta, in particular, who worked as teaching assistants in the IPR course and asked me insightful questions while grading the exams. This helped me to understand the needs of students better.

I thank Professor Prasad, who invited me to conduct the GIAN Course on IPRs to law and economics professionals at IIT-K. This helped me to learn about the needs of working professionals with diverse interests in IPRs.

Finally, I thank my wife Ramani, and my dear family—Maruti, Mayur, Manoj, and Vandana—who gave me incredible love, encouragement and support to complete this book.

Uday S. Racherla
Fort Lee, New Jersey

Author

Dr. Uday S. Racherla served as the *Professor of Innovation and Intellectual Property Management* in the Industrial Engineering and Management Department at the Indian Institute of Technology Kanpur, India, during 2012–2017, before returning to the USA. Prior to this, he taught briefly at the Carey School of Business, Johns Hopkins University, Baltimore, Maryland. Professor Racherla's academic areas of interest include Strategy, Innovation & Entrepreneurship, Technology Management, Intellectual Property Management and Sustainability.

Prior to the academic stint, Dr. Racherla had 20+ years of global innovation leadership and R&D management experience in Fortune 500 companies in the USA. During this period, Dr. Racherla held many top executive positions—including, Senior Director of Innovation at PepsiCo, Worldwide Director of R&D at S. C. Johnson, Director of R&D at New Skin Enterprises, and other senior executive positions at Unilever—and led global R&D programs.

Dr. Racherla invented, innovated, patented and launched many successful commercial products worldwide. His industrial R&D expertise included the areas of skin care, laundry detergents, home sanitization, air and water purification, and nutritious foods.

Dr. Racherla earned his PhD from Purdue University, Lafayette, Indiana, under a Nobel Laureate, post-doctoral research experience from The Ohio State University, Columbus, Ohio, under a distinguished professor of chemistry, and an executive MBA from the Kellogg School of Management, Evanston, Illinois. He published more than 120 research papers from academia and industry in prestigious journals such as *Nature* and *Journal of the American Chemical Society*, served as an editor of a book, authored 2 book chapters, graduated several PhD and MTech students, obtained 12 patents, designed and conducted multiple executive training courses, served as visiting professor at the Singapore Management University (SMU), and delivered many invited lectures at prestigious institutions worldwide.

Section I

Foundations

1

The Underpinnings of Intellectual Property Rights (IPRs)

> Think of the *information revolution* as one of the three great revolutions in the *cost of transport*. The nineteenth century, dominated by the steamship and the railway, saw a transformation in *the cost of transporting goods* and the twentieth century, with first the motorcar and then the airplane, in the *cost of transporting people*. The new century will be dominated by the transformation in the *cost of transporting knowledge and ideas.*
>
> **Frances Cairncross**

In this book, my goal is to present the subject of *intellectual property rights* (IPRs) from the unique perspectives of *business strategy, innovation,* and *technology management* to scientists, engineers, and business professionals in both academia and industry. Indeed, what this book will not do is to elaborate IPRs from a legal angle, as many scholarly works are already available for that purpose.

Therefore, before a detailed discussion on the various aspects of IPRs—such as, what are IPRs, how are they useful, who needs them, and how the international IP systems differ from each other, among other important things—we must get acquainted with the *underpinning concepts* and *ideas.*

IPRs—The Three Spaces of Expertise and Three Dimensions of Value

"IPRs protect[1a] the creations of mind, such as inventions; literary and artistic works; designs; and symbols, names, and images used in commerce." In the commercial world, IPRs involve three *spaces of expertise* and three *dimensions of value*: the "technology" space, wherein products/services are created based on new/improved technological *inventions,* to solve the needs/wants/challenges of various stakeholders (focused on *value creation*); the "business" space, wherein the (new/improved) products/services are wrapped in a *business model* and commercialized to capture the revenues/profits/market

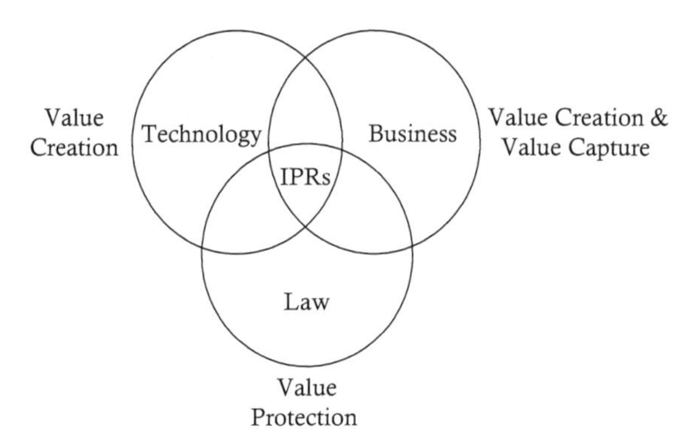

FIGURE 1.1
The three *spaces of expertise* and three *dimensions of value* of IPRs.

share by a firm (focused on *value creation*[1b] as well as *value capture*); and the "law" space, wherein IPRs are actually secured to protect *inventions* and *innovations* (focused on *value protection*).

Figure 1.1 summarizes these important concepts.

Tangible Assets and Intangible Assets of Firms

To start with, we must know what an "asset" is. From an accounting perspective, an "asset" may be defined as *an economic resource, owned by an individual or a firm that can be converted into cash.*[2] Such a definition allows us to distinguish two different kinds of assets, namely: (a) *tangible* assets (TAs) that are *objective and physical/financial in nature* (such as land, buildings, cars, furniture, jewelry, cash, stocks, bonds, etc.) and (b) *intangible* assets (ITAs) that are *identifiable, non-monetary,* and *non-physical.* It is important to note that there are many different *intangible* assets.[3,4] For the purpose of simplicity, however, we classify ITAs as: (1) legal intangibles (LIs); (2) trade secrets (TSs); (3) codified proprietary knowledge (CPK); and finally (4) competitive intangibles (CIs).

Interestingly, there are differences in ITAs among the intellectual property regimes of different nations. For example, in the United States of America (USA), the LIs include[5a]—*patents, copyrights,* and *trademarks* that are protected by the United States (U.S.) intellectual property laws. In India, the list of LIs include—in addition to what is listed for the USA—many other forms of ITAs, such as *industrial designs, integrated circuits, plant varieties, geographical indications,* and *traditional knowledge* that are protected by the Indian IP laws. In the USA, *trade secrets* are given protection under state laws (the 1979 Uniform Trade Secrets Act[5b]) and hence are treated separately from other LIs.[5c] In India, TSs are often listed as part of LIs, though they are not registered under a national organization, or granted by state, and do not expire until discovery or loss.[6]

Often, firms possess many forms of ITAs that cannot be captured as LIs. Those ITAs fall under the categories of CPK and CIs. A firm's CPK includes *all confidential information that is valuable and exclusive property of the company*, such as—standard operating procedure manuals, internal reports that capture their specialist know-how, in-licensing/out-licensing agreements, franchisee contracts, strategic alliances, research note books in which secret formulae and unique compositions are noted, and other difficult-to-imitate business/technology processes, among others. A firm's CIs include *valuable, difficult-to-quantify ITAs, such as firm's culture, innovative abilities, core competencies, unique buyer/supplier relationships, brand equity, reputation*, and *goodwill*, to mention a few. In fact, due to this reason, the *valuation of a firm* is often very difficult, as there can be many ITAs that a successful firm may possess—that are important sources of its competitive advantage and overall value—which defy straightforward valuation methods. Thus, while we have noted that a firm's ITAs can be many, they can all be organized into two groups of assets, namely—*Intellectual Assets* (IAs) and CIs. This leads us to the question: What is the relationship between IAs and IPRs?

Intellectual Assets and Intellectual Property Rights

A firm's IAs are a sum of the company's IPRs, TSs, and CPK. Thus, IAs are more than the IPRs. Further, a firm's ITAs are a holistic sum of the company's IAs and CIs. Figure 1.2 summarizes these details.

A firm's IPRs are its LIs. However, as already mentioned, IPRs vary depending on the IP regimes of different countries. To summarize, in America, IPRs include— *patents, copyrights*, and *trademarks*, but they are very broad in scope of protection of IP.[5] In China, IPRs include—*patents, copyrights, trademarks*,

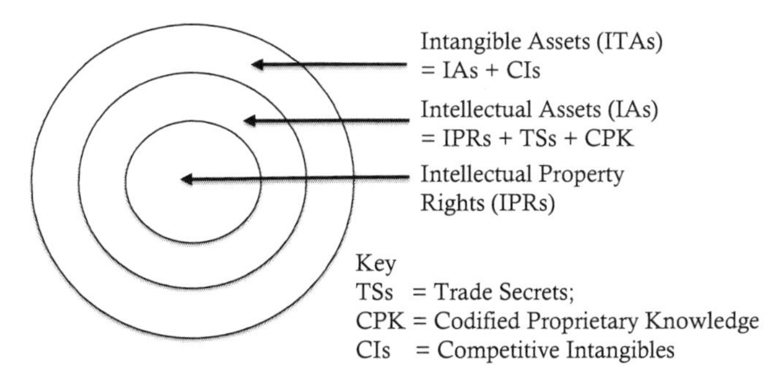

FIGURE 1.2
The relationship between various intangible assets (ITAs).

and *designs.*[7] In the European Union (EU), IPRs include more forms than in China, namely—*patents, copyrights, trademarks, industrial designs, new varieties of plants,* and *geographical indications of origin.*[8] In India, there are even more forms of IPRs, such as—*patents, copyrights, trademarks, trade secrets, industrial designs, integrated circuits, plant varieties, geographical indications,* and *traditional knowledge.*[6,9]

Net Worth/Book Value, Market Capitalization, and True Market Value

In the commercial world, companies are under continual pressure by investors and shareholders to achieve sustainable profitability and growth. As a result, companies are continually balancing their business strategies[1] of: (a) *organic growth* involving *innovation* of new and/improved, products/services/technologies and (b) *inorganic growth* based on acquisitions/mergers/strategic alliances. Thus, *organic growth* requires *innovation* and *technology management.* On the other hand, *inorganic growth* requires a clear understanding of the *net worth* of the targeted companies for acquisition/merger and expertise in *valuation* and *business management.* All of this requires a grasp of *what intellectual capital (IC) management is,* and *how it defines the value of the company.* To understand this, we need to know—the *net worth/book value, market capitalization,* and *true market value* of a firm.

Net Worth/Book Value[10a,b]

"Net worth" of a firm is, a "quantitative measure of the firm's worth as an entity." To determine the net worth of a firm, we must quantify the value of all its TAs, ITAs, as well as liabilities. In the field of accounting, there are reliable, accurate and widely accepted rules and methods for valuation of *tangible* assets/liabilities. In fact, companies routinely report the value of their TAs (physical and financial assets) on their balance sheets. However, the challenge lies with the valuation of ITAs. In the accounting world, *net worth* is also known as the "book value" or *shareholder's equity.*

"Book value" of a firm may be defined as, *the value of total assets of the company (tangible* and *intangible* assets*) minus its total liabilities.* It is calculated based on the data reported in the balance sheet on assets and liabilities. In today's *knowledge economy,* companies invest far more in ITAs (such as IPRs, TSs, CPK, and CIs), than in *tangible* assets (such as property, plant, and equipment). In the present system of accounting rules, however, companies don't report all their ITAs on their books (NOTE: some companies report the

[1] These must be aligned with company's vision, mission, and values.

value of *patents* and goodwill). Moreover, valuation of the ITAs of a firm—such as all IPRs, TSs, CPK, and CIs—is not simple, precise, and foolproof. Consequently, a firm's investments in ITAs often remain almost completely hidden from the investors. As a result, three scenarios may result: (a) if a firm's ITAs go *unrecognized* and *underrepresented* on the balance sheet, *as is usually the case,* then the *book value* of the firm will be *significantly lower than its net worth;* (b) if a firm's ITAs are *improperly undervalued,* then also the *book value* of a firm will be *lower* than its *net worth,* and, finally (c) if a firm's ITAs are *improperly overvalued,* then the *book value* of the firm will be *higher* than its *net worth.* In essence, the *book value* of a firm *may not be a reliable indicator* of its *net worth.* For this reason, the investment community often looks to other indicators for the net worth of firms, such as—*market capitalization* and *true market value.*

Market Capitalization and True Market Value[11]

The "market capitalization" of a company may be defined as, *the current price of a single stock multiplied by the total number of outstanding shares of the company.* Clearly, *market capitalization* is a reliable and superior quantitative measure of the *net worth* of the company compared to the *book value.* Yet, even this snapshot picture of a firm is not a perfect measure of its *true market value.* The *true market value* of a firm is even more complex—as it involves numerous metrics, such as *price-to-earnings, price-to-sales,* and *return-on-equity*—to capture *the growth potential* of the firm. Thus, the *true market value* of a firm accounts for many other factors such as—outstanding bonds, long-term growth potential, company debt, taxes, and interest, besides shareholder's equity. Therefore, quantifying the *true market value* of a firm is even more challenging. Thus, from the above discussion, it is hopefully evident that there are no simple methods for pinning down a firm's *true market value,* as it requires accurate valuation of all its ITAs and their impact on growth—a combination of *objectivity* and *subjectivity.*

Information Revolution and Knowledge Economy

Individuals, organizations, societies, and ultimately nations create *competitive advantage* and achieve *socio-economic progress based on their ability to rapidly share information and excel in scientific discoveries, inventions, and innovations* (both technological and business model) that can effectively solve the needs/wants/challenges of the world. Indeed, *inventions* and *innovations* have the power *to transform the whole world.* Thus, innovation of the printing press led to mass literacy, promoted the "Protestant Reformation" (by undermining the clergy's theological monopoly), and enabled the scientific revolution—all because of the *information revolution* it started. Similarly, major innovations in shipping, railways, automobiles, and aviation paved the way for rapid and efficient transportation of people and cargo across long distances, *to trigger a worldwide revolution in trade and commerce*. By the same token, the invention of telephone, telegraph, television, and satellite-based communications paved the way for a *long-distance communication revolution*, making time-consuming information exchange by other means obsolete, and advanced the cause of *information revolution* even further. Similarly, the birth of mainframe computers and computer languages enabled the *rapid computation* and *complex problem-solving possible*, kicking the *information revolution* to the next higher level.

Today, the *digital information revolution* made possible by the innovations of the Internet, computer hardware (such as desktops, laptops, tablets, cell phones, smartwatches, etc.), social media (such as YouTube, Facebook, LinkedIn, and Twitter), and cloud computing—are making communication and exchange of information possible, like never before in the history of the world. Thus, the instant availability of vast amounts of data and complex information at fingertips, coupled with rapid means to use/communicate/evaluate that information across the continents of the world without any travel, is *positively transforming the way we think, live, and work*. For example, an e-commerce company (such as Amazon) may be located on one continent, but its businesses (products and customers) are global. Similarly, a person may reside on one continent, but work for a company on a different continent. Consequently, in today's world, work goes wherever it can be done effectively, efficiently, and at low cost, regardless of the country or continent. Thus, it is these factors that have created the *knowledge economy* and *globalization*.[12b]

As a result, *knowledge economy* has increased the contribution of *intangible* assets to the market value of firms in the last four decades. Indeed, the 2015 annual study of the Intangible Asset Market Value of S&P 500 Companies by Ocean Tomo[13] highlighted the *exceptional growth in the contribution of intangible assets to the market value of S&P 500 firms* over a period of 1975–2015 (Table 1.1).

Therefore, according to Sveiby,[14] *stock price* is the only reasonable measure for quantifying the *market value* of a firm's ITAs.

Thus, in *knowledge economy*, the *stock prices* of companies richer in ITAs are found to be much higher than the *stock prices* of companies with

TABLE 1.1

The Growth of Intangible Asset Market Value of S&P 500 Companies

	Contributions to Market Value				
Asset Type	**1975 (%)**	**1985 (%)**	**1995 (%)**	**2005 (%)**	**2015 (%)**
Intangible assets	17	32	68	80	84
Tangible assets	83	68	32	20	16

Source: Retrieved on September 18, 2016 from: http://www.oceantomo.com/blog/2015/03-05-ocean-tomo-2015-intangible-asset-market-value/.

abundant TAs. In fact, Sveiby's[14] research revealed that the *stock prices* of pharmaceutical companies are valued several times higher than the service-based companies, which, in turn, are higher than the real estate companies whose assets are primarily *tangible*. Therefore, the question is: "Are there alternative approaches to quantify the value of all ITAs of a firm, so we can arrive at its *true market value*?" To answer this question, we must know *Intellectual Capital* (IC).

Intellectual Capital and Its Importance in Knowledge Economy

In 1868, the "balance sheet" was born in its present form. The "balance sheet" coupled with "income statement" systematized the reporting of the financial results of companies. According to Elliott,[15] "[the balance sheet] focuses on tangible assets, that is, the assets of the industrial revolution. These include inventory and fixed assets: for example, coal, iron, and steam engines. And these assets are stated at cost. Accordingly, we focus on costs, which is the production side, rather than the value created, which is the customer side." Thus, in early industrial economy, when *tangible* assets (physical and financial) *alone* were sufficient for valuation of companies, the *book value* of a company provided a fairly reasonable estimate of its *net worth* and *market value*.

However, in *knowledge economy*, when ITAs began to account for 80%–90% of the *market value* of the companies and, when the *standard accounting procedures* became unreliable for estimating the *net value* of ITAs, the need for a *different theoretical basis for valuation of companies* became obvious. Thus, Edvinsson[16] created a new concept known as the IC and the following equation for *market value*:

Market Value = Book Value + Intellectual Capital (IC)

We can now ask: "What is IC?" Stewart, a pioneer in the study of ITAs, defined IC as,[17a]

> the intellectual material—knowledge, information, intellectual property, experience—that can be put to use to create wealth; the intellectual material that has been formalized, captured, and leveraged to create wealth by producing a higher-valued asset; it is packaged useful knowledge.

Further, Stewart[17a] explained IC as, when the stock market values companies at three, four, or ten times the *book value* of their assets, it's telling a simple, but profound, truth: the TAs of a knowledge company contribute far less to the value of its ultimate product (or service) than the ITAs—the talents of its people, the efficacy of its management systems, and the character of its relationships to its customers—that together constitute its IC.[17b] In *knowledge economy,* all ITAs are collectively referred to as *knowledge assets* (KAs), and their management is known as *knowledge management.* To start with, it is important to note that *knowledge management* attempts to identify, formalize, and capture all the *intellectual material* of the company—tacit knowledge, information, expertise, and intellectual property—into assets of higher value, which can then be leveraged to create value.

Therefore: "What is *IC Management*?" While a company may have multiple ITAs (or KAs), *knowing those which will be specifically valuable to its business in the present and future—based on own vision, mission, and strategy (strategic fit)—is vital.* Such KAs (or ITAs) constitute company's IC. Once the company knows its IC, then it can manage its IC to create and capture value for the company. Firstly, all the KAs that show clear-cut *strategic fit* will be leveraged directly in company's business to create competitive advantage. Secondly, among the remaining KAs, the company may harvest a few by out-licensing/selling *to generate an alternative stream of wealth.* Thirdly, the company may employ some of its KAs as collateral *to raise capital for its business* or use some KAs for *cross-licensing purposes.* Finally, the company may donate some of its KAs to others (non-competitors) *to build positive reputation and goodwill* and even abandon some KAs *to avoid the costs of maintenance.* Thus, *knowledge management*[18] involves the identification, formalization, and capture of all KAs (or ITAs), while *IC management* involves the leverage of all KAs (or ITAs) into creation and capture of value.

Though there are many viewpoints, there is a consensus that IC *is composed of three elements,* namely—human capital (HC), structural capital (SC), and relational (customer) capital (RC).

According to Stewart, "Human Capital" is "the well-spring of *innovation* and the home page of *insight.*"[17c] This view is based on the idea that—money is critical for the company to run, but it is the *people* who make *creative thinking possible;* machines are critical for production, but it is only *people* who can *invent* and *innovate.* Thus, HC is *the knowledge, skills, and capabilities of individual employees of a company needed to create innovative solutions (products/services) to address the needs/wants/challenges of various stakeholders (suppliers, vendors, buyers, consumers, creditors, shareholders, and investors).* It is important to point out that HC can be *only rented, but not owned, by a company,* as it enters the company when employees come to work in the morning and leaves with them when they go home in the evening. Consequently, the HC of a company *can't be taken for granted, as it needs inspiration, motivation, nurture, recognition, and appropriate reward to create innovations that provide sustainable competitive advantage.*

Next, Structural Capital is a *company's* IAs—IPRs, TSs, and CPK (such as *confidential data bases, company reports, SOP manuals, licensing agreements, business contracts,* etc.)—*coupled with business systems* (operational and decision making), *that enable the organization to create value in a repeatable, scalable way.* In other words, SC is "captured and institutionalized knowledge and capabilities." Unlike HC, SC is *owned by the company,* and it does not leave the organization when employees leave for home in the evening or join a different company.

Building on Stewart's definition[17d] of "Customer Capital," "Relational Capital" (RC) can be understood as *the value of a company's relationships with consumers, vendors, suppliers, buyers, strategic partners, industry associations, Non-Governmental Organizations (NGOs), quality assurance, regulatory agencies,* and *policy makers—which contribute to business success.*[19] According to Saint-Onge,[20] "customer capital" *is the depth (penetration), width (coverage), and attachment (loyalty) of a franchise.* From now on, we will refer to this concept only as RC. In RC, we measure and quantify how firms *integrate* and *leverage external relationships with all their customers* (loyalty, trust, goodwill, and reputation) *into sustainable value.* Clearly, RC *cannot be owned or controlled by a company,* but *can only be nurtured and managed* by it, as it is suffused in all its relationships with its customers, upstream or downstream in the value chain. In some companies, RC is measured and often compared to HC or SC. For example, a company looking for *category expansion/geographical expansion* of its products/services,[21] will often engage in measurements of *brand equity*—which is a form of RC—as it could reveal the full scope of company's product/service offerings. The answer to the question: *Is IC management critical to firms?* is, "Absolutely." Extending Serrat's logic,[22] IC management involves three objectives:

- *Value creation*—which involves proactive creation of new IC;
- *Value extraction*—which involves leveraging of existing IC into value; and
- *Value reporting*—which involves valuation and timely reporting of IC to all stakeholders, to enhance the firm's reputation and *market value.*

Relationship of Intellectual Capital and IPRs

It is important to note that IPRs are an integral component of the SC of the company, which in turn is a key component of the IC of the firm. Therefore, technically speaking, *IP management* is included in *IC management.* However, as we will show later in Section III of this book, the details of *strategic IP management,* in their own right, are crucial as they require knowledge and expertise in strategies (defensive, offensive, and value enhancing) that decide whether a company gains/loses its *competitive advantage* in the market place and assess whether a company's innovations can create sustainable value to all stakeholders.

Value of IPRs—Promotion or Obstruction of Innovation?

There has been a perennial debate on *whether IPRs help/hurt creativity* and *innovation.* At the center of the debate are two opposing ideas: *exclusivity* vs. *open access.* Unfortunately, perspectives in this debate get sometimes muddled. It is important to clarify that IPRs award only *limited exclusivity* to inventors, writers, and artists for public disclosure of their inventions, literary and artistic works, symbols, names, images, and designs of commercial value. No one gets away with *unlimited exclusivity.*

The proponents of IPRs argue that IPRs help *creativity* and *innovation* and cite considerable evidence in favor of that argument.[23a] Indeed, this is the rationale for the following declaration by the World Intellectual Property Organization[23b]:

> At World Intellectual Property Organization (WIPO), we believe that IP and *innovation* go hand-in-hand. IP is about rewarding people for their ideas and original creations. It is about encouraging businesses to invest in *innovations* and new solutions. It is a means of enabling *innovation* and *creativity* for the benefit of all.

On the contrary, the opponents of IPRs also cite what they believe is valid evidence and argue that IPRs hurt *unfettered access* to knowledge and creativity.

This book points out the arguments on both sides. However, this author agrees with the views expressed by IPR proponents, based on the evidence and facts presented in the ESO-USPTO study and OHIM-EPO study (discussed below), and author's own expertise in invention, innovation, and commercialization in Fortune 500 companies.

Perspectives of IPR Opponents

Building on the narrative by Gollin,[24] IPR opponents cite at least *eight arguments* against IPRs:

- **IPRs obstruct unencumbered public access to innovation.** *When exclusive rights are granted in an excessive manner" (either in length of term or breadth of coverage), then that obstructs the full open access to innovation. Thus, the IPR holder receives asymmetrically higher benefits than the society at large.* But, there is a flaw in this argument. Exclusive rights are given only for a limited period of time to recover the costs and incentivize further innovation. Also, who decides "what the acceptable" length/breadth of coverage is? Further, unfettered public access to innovations is also a flawed logic. Indeed, it is unfair to argue that the hard-earned outcome of one's creative work must be free to all and not decided by the free market forces.

- **IPRs lead to higher costs to consumers.** *When an inventor, artist, or author receives exclusive rights, then those rights are typically used to charge higher prices to consumers.* True, a pharmaceutical company may charge exorbitant prices for its new "orphan drug,"[2] putting them out of reach of low-income consumers. However, prices of all goods and services are always determined by the free market forces of supply and demand. In special cases, governments may step in to make medicines affordable to low-income consumers through price control regulations or subsidies.
- **IPRs lead to creation of monopolies.** *IPRs create unhealthy monopolies by certain companies, in specific technologies, within some geographic regions.* For instance, this line of argument goes as follows: suppose, a pharmaceutical company brings new innovations, on a continual basis, into a country with high incidence of specific human ailment, such as heart disease, it will gain monopoly in that market. Again, this is a false argument. No company can prevent another company from innovating equally powerful or superior drugs based on alternative mechanisms, for a specific human disease. Furthermore, monopolies in the market place should be prevented by the governments through market scrutiny, regulations (example, anti-trust laws), and corruption-free law enforcement.
- **IPRs lead to a focus on profit rather than on service.** This perspective arises from the notion that *companies focus only on innovations (products/services) that are protectable by IPRs and not on innovations that may not be protectable by IPRs, though they may render superior products/services.* Indeed, this is contrary to facts, as any industry practitioner will vouch for. Sure, every company may wish that all its innovations are fully protected by IPRs. Firstly, this will be impossible even for the greatest company in the world! Secondly, no company "gives up" innovations that offer at least a short-term *competitive advantage, just because they are unprotectable by* IPRs. In fact, the goal of companies that follow the value discipline[25] of *product leadership* (such as Procter & Gamble, Apple, Google, etc.) is to continually launch products/services to be either *first-to-market* (to stay ahead of competition) or *best-in-the-market* (to be always better than competition). Here, the *speed* and *quality of innovations* are the focal points for companies. This has *everything to do with both profitability and service*! Of course, companies would love if all those innovations are protected by IPRs.

[2] Orphan drugs command hefty prices in the market, as they target rare diseases. In these cases, the low volume of prescriptions is countered by high unit prices to ensure the desired returns on investment.

- **IPRs lead to competition rather than cooperation**. According to this argument, *inventors conceal their knowledge, expertise, and valuable findings from the competition in particular, and society in general, until they protect them with IPRs*. Sure, but: "What is wrong with that?" Consider the following example. Suppose, you are a company (inventor) that invented a new light bulb that can disrupt the market—because it runs on solar energy and costs very little to make—after years of research and spending tens of millions of dollars and you have not yet created a product/service for commercialization, why should you hurry to publicize it to everyone, even before your invention is protected? How can anyone find fault with a company for safeguarding its investment and commercial interests? How can cooperation (prematurely publicizing findings, and valuable expertise) prior to IP protection, let alone commercialization, be fair to the inventor and beneficial to company or its investors? In fact, while IPRs guarantee exclusive rights, *they also incentivize inventors to disclose inventions and bring their knowledge into the public domain.*
- **IPRs are costly to acquire and retain**. According to this school of thinking, *not all individuals and organizations can afford to pay for acquisition/retention of IPRs, so the class of inventive poor is unfairly discriminated.* Clearly, this argument is about fairness. While everyone agrees that the cost of acquisition and maintenance of IPRs should be brought within the reach of everyone, it is also true that no nation on earth has yet been able to eliminate poverty or social injustice.
- **IPRs necessitate dealing with complex legal and regulatory framework**. *IPRs necessitate dealing with elaborate legal and regulatory framework to comply with various international conventions and treaties* (these will be described later in the book), *which may intimidate some and incur legal costs that all may not be able to afford.* Once again, *the spirit of this argument appropriately highlights the need to protect the interests of the inventive poor.*
  - Indeed, this situation must be addressed by other means: by creating free IPR education programs and pro bono services to the inventive poor. In fact, one such international non-profit organization, Public Interest Intellectual Property Advisors founded by Frank Gollin in Washington, DC, provides *pro bono intellectual property (IP) legal counsel to governments, businesses, indigenous peoples, and public interest organizations in developing countries that seek to promote health, agriculture, biodiversity, science, culture, and the environment.*[26]
- **IPRs are unethical**. Differing perspectives[24] on IPRs stem from the inherent tensions present in conflicting ideologies (political, social, and economic) and their associations, such as: rich vs. poor;

powerful vs. weak; private vs. public; closed vs. open; slavery vs. freedom; capitalism vs. socialism; individuality vs. collectivism; us vs. them; moral vs. immoral; and finally, right vs. wrong.

Some critics carry this kind of reasoning to the extreme and ask: *Could you make the argument that IPR laws themselves are immoral or unethical in that they are actually what makes everyone worse off?*[27] We will not go there.

Perspectives of IPR Advocates

There are *several arguments* that the *IPR advocates* make in favor of IPRs.

- **IPRs incentivize creativity**. The central idea here is that *creativity is incentivized by reward*. Thus, *if an individual gains* IPRs (for a limited time, of course) *for his/her intellectual creations*—for investing in imagination, creativity, time, effort, and money to: (a) invent a product that solves a (complex) problem, (b) write a book of transformational ideas on an important subject, or (c) create a unique design for an iconic building—*and is able to reap the economic benefits and/or social recognition for the same* (rewards), then this incentivizes further investment of time and effort in additional creative pursuits. Indeed, this is also true for *organizations* and even *nations* engaged in *creativity* and *innovation*.
- **IPRs encourage public disclosure of information**. *If IPR protection is afforded to inventors, authors, and artists for the intellectual creations, in exchange for public disclosure, then they will be incentivized to share their knowledge and expertise with others*. Without IPRs, anyone can unethically exploit the creative labor of others (inventors, authors, and artists), and this will breed a predatory environment.
- **IPRs strengthen the businesses and improve the business environment**. *IPRs make businesses stronger and introduce a healthy competition into the industry*—to constructively improve the business environment, in many ways:
    - IPRs enable companies to achieve IP-protected *differentiation of their offerings* (products/services);
    - IPRs empower companies to make *stronger market claims* for its offerings;
    - IPRs can serve as collateral to *raise financial capital;*
    - IPRs encourage growth-seeking companies to do *strategic IP management,* for example, conduct annual IP audits to decide which IP must be—kept (to grow the company according to its vision, mission, and strategy), licensed/sold (to generate an alternative stream of revenues), donated (to gain relational capital), or even abandoned (to avoid unnecessary maintenance costs);

- IPRs challenge existing rivals in an industry to *earn their competitive advantage* through *strategy, creativity*, and *innovation*; and
- IPRs force competitors to continually invent/improve their technologies to sustain profitability or face extinction.

- **IPRs increase the valuation of companies.** As already mentioned, in today's *knowledge economy*, ITAs account for more than 80% of the *market value* of a firm (Table 1.1). Therefore, *the value of a company possessing strong IPRs is significantly higher than otherwise* (see below).
- **IPRs promote economic growth and create employment.** There have been many studies on the role of IPRs in promoting economic growth.[28]

  According to a recent study by the European Office for Harmonization in the Internal Market (OHIM)[29]:

  > The economic well-being of the EU relies on sustained creativity and innovation. The Europe 2020 strategy has defined smart growth as one of the European Commission's three policy priorities for the future to create a more competitive economy with higher employment. In this respect, IPRs have been identified as instrumental for the Union's future prosperity, innovation capacity, and growth policies.

  This study also reported that 96% of EU citizens agree that inventors, creators, and performing artists should be able to protect their work and get paid for it; similarly, 86% of EU citizens opine that IPRs contribute to improving/guaranteeing the quality of products/services. However, IPR opponents demand a clear evidence that proves the critical role of IPRs in building economies and providing employment. Accordingly, there have been two systematic studies reported—one based in the USA and another based in the European Union—that unequivocally show the relationship between IPRs, economic growth, and employment. They are:

  - The Economics and Statistics Administration (ESA) - United States Patent and Trademark Office (USPTO) Study[30]; and
  - The Office for Harmonization in the Internal Market (OHIM) - The European Patent Office (EPO) Study.[31]

The ESA-USPTO Study[30]

Firstly, this investigation identified the most IP-*intensive* industries in the U.S. economy (whom we define as the "initiators"), based on several industry-level metrics on IP that it developed. For this purpose, the study used several databases, some of which (for the patent and trademark analyses) are in the public domain.[32] Secondly, this study examined IP initiators that: (a) protect their innovations through *patents, trademarks,* or *copyrights*; (b) contribute several trillion dollars to the U.S. Gross Domestic Product (GDP); and (c) create/support hundreds of millions of jobs. Thirdly, this research estimated the secondary effects of initiators on the U.S. economy and employment, as the initiators stand for ingenuity and exemplify the fact that the growth of a firm/nation is dependent on the effective protection of IP rights.

The following are some of the major findings of the study:

- **IP dependence**. The U.S. economy as a whole depends on IP, as virtually every industry is either a *creator* or *user* of IP.
- **Employment**. This study focused on 313 U.S. industries and identified 75 of them as *IP-intensive*. In 2010, these 75 *IP-intensive* industries (initiators) directly accounted for 27,100,000 jobs, or 18.8% of all jobs in the U.S. economy. A substantial share of this *IP-intensive* employment is due to 60 *trademark-intensive* industries, which accounted for 22,600,000 jobs. In addition, 26 *patent-intensive* industries accounted for 3,900,000 jobs, while 13 *copyright-intensive* industries provided 5,100,000 jobs, during the same year.
    - While 75 *IP-intensive* industries (initiators) *directly* accounted for 27,100,000 jobs, they also *indirectly* supported 12,900,000 additional supply chain jobs in 2010. Stated differently, two jobs in an *IP-intensive* industry supported an additional job elsewhere in the economy. Thus, 27.7% of all the U.S. employment (40,000,000 jobs) is *directly/indirectly* attributable to initiators.
    - *Globalization,* the rise of emerging economies, the availability of cheaper labor, and the advent of free trade agreements between nations—have led to job losses in the U.S. manufacturing sector, as well as a slowdown in the overall employment in *IP-intensive* industries, during the last two decades. Thus, during 1990–2011, employment in *non-IP-intensive* industries grew by 21.7%, while employment in the *IP-intensive* industries grew only by 2.3%. Also, *patent-intensive* industries experienced relatively higher job losses, especially during 2000–2010, emblematic of the situation of the manufacturing sector in the USA. Furthermore, while employment in the *trademark-intensive* industries declined by 2.3%, *copyright-intensive* industries showed a sizeable employment boost, growing by 46.3% over the same period (1990–2011).
    - In 2010–2011, economic recovery in the U.S. resulted in improvement of *direct employment* in *IP-intensive* industries by 1.6%, slightly higher than the growth in *non-IP-intensive* industries (1%). Growth in *copyright-intensive* industries (2.4%), *patent-intensive* industries (2.3%), and *trademark-intensive* industries (1.1%)—all outpaced the gains in *non-IP-intensive* industries.

- **Wage premium**. Another notable finding of the study was that *IP-intensive* industries paid higher wages compared to *non-IP-intensive* industries. Thus, in 2010, the average weekly wages (AWW) in *IP-intensive* industries ($1,156) were 42% higher compared to *non-IP-intensive* industries ($815). Further, this wage premium in *IP-intensive* industries increased from 22% to 42%, during 1990–2010. Specifically, *patent-* and *copyright-intensive* industries experienced a relatively rapid growth in AWW. In fact, the wage premium in *patent-intensive* industries increased from 66% to 73% in five years (2005–2010), while the same in *copyright-intensive* industries showed an increase from 65% to 77%.
- **Education level and compensation**. Analysis of college educated workers of age >25 revealed that they constituted >42% of workers in *IP-intensive* industries, compared to (on average) 34% in *non-IP-intensive* industries. Further, the comparatively high wages received by workers in *IP-intensive* industries vs. *non-IP-intensive* industries, corresponded to, on average, the higher levels of education of these workers.
- **Export contribution**. In 2010, the exports of merchandise in *IP-intensive* industries accounted for 60.7% ($775 billion) of total U.S. merchandise exports.
- **Foreign trade**. This report found that in 2007, approximately 19% of the total U.S. private service exports is accounted for by *IP-intensive* service-providing industries.
- **GDP contribution**. In 2010, *IP-intensive* industries (initiators) contributed $5.06 trillion (or 34.8%) to the GDP.

Thus, the ESA-USPTO research study reached the following conclusion:

> The granting and protection of IPRs is vital to *promoting innovation* and *creativity* and is an essential element of our free enterprise, market-based system. *Patents, trademarks,* and *copyrights* are the principal means used to establish ownership of inventions and creative ideas in their various forms, providing a legal foundation to generate tangible benefits from *innovation* for companies, workers, and consumers. Without this framework, the creators of *intellectual property* would *tend to lose the economic fruits of their own work, thereby undermining the incentives to undertake the investments necessary to develop the IP in the first place.*

The OHIM-EPO Study[31]

Just like the ESA-USPTO study, the OHIM-EPO study is the foremost investigation to systematically measure *the impact of IPR-intensive industries in the EU economy in terms of output, employment, wages, and trade* based on key forms of IPRs, namely—*patents, trademarks, designs, copyrights,* and *geographical indications.* (NOTE: the *IPR-intensive* industries in this study are same as the *IP-intensive* industries in the above study).

The purpose of this study was to establish factual evidence on the role and impact of IPRs in the EU economy and communicate this valuable information to policymakers, IP offices, business groups, and academics. This study defines *IPR-intensive* industries as *those having an above-average use of IPR per employee* and, on that basis, notes that 50% of the EU industries are *IPR-intensive.* The study points out that while every industry uses IPRs to some extent or the other, the fact that it focuses only on the *IPR-intensive* industries may underrepresent the impact of IPRs to the EU economy.

The major findings of this study are as follows:

- **Employment**. In the EU, *IPR-intensive* industries accounted for nearly 26% of all the jobs during 2008–2010. Further analysis revealed that the jobs created were spread over a wide range of *IPR-intensive* industries, such as *trademark-intensive* industries (21%), *design-intensive* industries (12%), *patent-intensive* industries (10%), *copyright-intensive* industries (3.2%), and *GI-intensive industries* (0.2%).
    - During 2008–2010, about 56,500,000 jobs out of approximately 218,000,000 were directly created by the *IPR-intensive* industries. In addition, another 20,000,000 jobs were indirectly created in the supplier industries that provided raw materials and services to the *IPR-intensive* industries. Consequently, considering direct and indirect employment, about 77,000,000 (35.1%) jobs were attributable to the *IP-intensive* industries and its business partners.
- **Value added per employee and wage premium**. This study found that the value added per employee is higher in *IPR-intensive* industries than in the other industries. Based on economic theory, therefore, one can expect *IPR-intensive* industries to pay their workers higher wages than the *non-IPR-intensive* industries. Indeed, this study established that the AWW in *IPR-intensive* industries were 41% higher (€715) compared to AWW (€507) in *non-IPR-intensive* industries. This "wage premium" is 31% in *design-intensive* industries, 42% in *trademark-intensive* industries, 46% in *GI-intensive* industries, 64% in *patent-intensive* industries, and 69% in *copyright-intensive* industries.
- **Trade, exports, and imports**. In continuation, this study examined the role played by *IPR-intensive* industries in the EU's external trade. Accordingly, research found that *IPR-intensive* industries accounted for a majority of the EU's trade. This is because even commodity industries that trade well—such as the energy industry—are *IPR-intensive.* As a result, *IPR-intensive* industries account for 88% of the EU's imports, and 90% of the EU's exports. The trade deficit of the EU as a whole was approximately €174 billion (or 1.4% of GDP). Since

the *IPR-intensive* industries account for a higher share of the EU's exports than the EU's imports, they make a net positive contribution to the EU's trade position. Indeed, the EU's trade surpluses in *copyright-intensive, design-intensive,* and *GI-intensive* industries offset the trade deficit in *trademark-intensive* and *patent-intensive* products.

- **GDP contribution.** During 2008–2010, *IPR-intensive* industries contributed approximately €4.7 trillion (39%) to the EU's GDP. This is spread over many *IP-intensive* industries, such as *trademark-intensive* industries (34%), *design-intensive* industries (13%), *patent-intensive* industries (14%), *copyright-intensive* industries (4.2%), and *GI-intensive* industries (0.1%).

Indeed, this study is valuable because it assesses the joint contribution of all *IP-intensive* industries to the economies of the EU, as well as to the individual member states, in a systematic and reliable manner.

Finally, the OHIM-EPO study[32] conveys the following message:

> *Innovation* is one of the areas covered by the five key targets set in "Europe 2020," the ten-year growth strategy adopted by the European Union with a view to creating a more competitive economy with higher employment. It has never been so important to foster the "virtuous circle" leading from Research and Development investment to jobs—via innovation, competitive advantage, and economic success—as in today's world of increasingly globalized markets and the *knowledge economy.* This process depends on several different factors, but an efficient system of IPRs undoubtedly ranks among the most important, given IP's capacity to encourage *creativity* and *innovation,* in all its various forms, throughout the economy.

References

1. (a) Retrieved on September 15, 2016 from: http://www.wipo.int/about-ip/en/; (b) In the Business Domain, *value creation* happens due to *Business Innovations.*
2. Sullivan, A.; and Sheffrin, S. M. (2003). *Economics: Principles in Action.* Upper Saddle River, NJ: Pearson Prentice Hall.
3. https://www.grantthornton.ie/globalassets/1.-member-firms/ireland/insights/publications/grant-thornton-intangible-assets-in-a-business-combination.pdf
4. Austin, L. (2007). Accounting for intangible assets. *Business Review.* 9(1): 63–72.
5. (a) http://www.uspto.gov/. (b) Uniform Trade Secrets Act with its 1985 Amendments. (1985). Prefatory note and comments. *National Conference of Commissioners on Uniform State Laws. Annual Conference Meeting,* Minneapolis, MN. (c) In USA, Trade Secrets received protection under state laws which varied from state to state, until the enactment of the 1979 Uniform Trade Secrets Act.

6. Rahmani, Z. M.; and Rahman, F. (2011). Intellection of trade secret and innovation laws in India. *Journal of Intellectual Property Rights*. 16(4): 341, 347. Retrieved on September 10, 2016 from: http://nopr.niscair.res.in/bitstream/123456789/12449/1/IJPR%2016(4)%20341-350.pdf
7. Intellectual Property Office. Intellectual property rights in China. Retrieved on September 16, 2016 from: file:///H:/Book%20Projects/Taylor%20&%20Francis%20-%20IPRs%20for%20Scientists%20&%20Engineers/Manuscript%20in%20Preparation/Chapter%201/References-Chapter%201/IPRs/IPRs%20in%20China.pdf
8. Intellectual Property Rights in EU. Retrieved on September 16, 2016 from: http://europa.eu/youreurope/business/start-grow/intellectual-property-rights/index_en.htm
9. Types of IPRs in India. Retrieved on September 18, 2016 from: http://www.makeinindia.com/policy/intellectual-property-facts
10. (a) Retrieved on September 17, 2016 from: http://www.fool.com/knowledge-center/for-wiki-how-to-calculate-the-net-worth-on-financi.aspx; (b) http://www.investopedia.com/terms/n/networth.asp
11. Retrieved on September 17, 2016 from: http://www.investopedia.com/ask/answers/122314/what-difference-between-market-capitalization-and-market-value.asp
12. (a) Cairncross, F. (2001). *The Death of Long Distance: How the Communications Revolution is Changing Our Lives*. Boston, MA: Harvard Business Press; (b) Torr, J. (Ed.). (2003). *Information Age*. Farmington Hills, MI: Green Haven Press.
13. http://www.oceantomo.com/intangible-asset-market-value-study/ (retrieved on August 1, 2018).
14. Retrieved on September 18, 2016 from: http://www.sveiby.com/articles/MarketValue.html
15. Elliott, R. K. (1992). The third wave breaks on the shores of accounting. *Accounting Horizons*. 6(2): 61–85.
16. Edvinsson, L. (2002). *Corporate Longitude*. London, UK: Prentice Hall.
17. (a) Stewart, T. A. (1997). *Intellectual Capital: The New Wealth of Organizations*. New York City: Double Day/Currency Publishers, p. X; (b) *ibid*. p. 55; (c) *ibid*. p. 86; (d) *ibid*. p. 77.
18. Kok, A. (2007). Intellectual capital management as part of knowledge management initiatives at institutions of higher learning. *Electronic Journal of Knowledge Management*. 5(2): 181–192.
19. CIC. (2003). *Modelo Intelleäus: Medición y Gestion del Capital Intelectual*. Madrid, Spain: CIC-IADE.
20. Saint-Onge, H. (1996). Tacit knowledge: The key to the strategic alignment of intellectual capital. *Strategy & Leadership*. 24: 10–13.
21. *Category expansion* refers to introduction new products/services into the existing brand (geography remains the same), whereas *geographical expansion* refers to introduction of the existing category of products/services (category remains the same) into different geographical locations.
22. Serrat, O. (2011). A primer on intellectual capital. *Knowledge Solutions*. 106: 1. Retrieved on September 21, 2016 from: https://www.adb.org/sites/default/files/publication/29207/primer-intellectual-capital.pdf/

23. (a) ICC. (2016). Helping business use intellectual property. Retrieved on September 30, 2016 from: http://www.iccwbo.org/Chamber-services/Chamber-resources/Helping-business-use-intellectual-property/; (b) WIPO. (2014). Making IP work. Retrieved on September 23, 2016 from: http://www.wipo.int/edocs/pubdocs/en/general/1060/wipo_pub_1060.pdf
24. Gollin, M. A. (2008). *Driving Innovation: Intellectual Property Strategies for a Dynamic World*. Cambridge, UK: Cambridge University Press.
25. Treacy, M.; and Wiersema, F. (1997). *The Discipline of Market Leaders: Choose Your Customers, Narrow Your Focus, and Dominate Your Market*. Reading, MA: Addison-Wesley.
26. PIIPA. (2016). Retrieved on September 30, 2016 from: http://piipa.org/index.php
27. Koepsell, D. (2016). Is intellectual property itself unethical? Retrieved on September 30, 2016 from: https://www.techdirt.com/articles/20100519/0404029486.shtml
28. Gould, D. M.; and Gruben, W. C. (1996). The role of intellectual property rights in economic growth. *Journal of Development Economics*. 48: 323–350.
29. The OHIM Study. (2013). The European citizens and intellectual property: Perception, awareness and behavior. Retrieved on October 1, 2016 from: https://euipo.europa.eu/ohimportal/documents/11370/80606/IP+perception+study
30. The ESA-USPTO Study. (2012). Intellectual property and the U.S. economy: Industries in focus. Retrieved on October 5, 2016 from: https://www.uspto.gov/sites/default/files/news/publications/IP_Report_March_2012.pdf
31. The OHIM-EPO Study. (2013). Intellectual property rights intensive industries: Contribution to economic performance and employment in the European Union. http://ec.europa.eu/internal_market/intellectual-property/docs/joint-report-epo-ohim-final-version_en.pdf
32. See, (a) www.uspto.gov/web/offices/ac/ido/oeip/taf/data/naics_conc/, and also (b) www.google.com/googlebooks/uspto-trademarks.html

2

Fundamentals of Intellectual Property Rights (IPRs)

Innovation—the process through which new ideas are generated and successfully introduced in the marketplace—is a primary driver of U.S. *economic growth* and *national competitiveness*. Likewise, U.S. companies' use of trademarks to distinguish their goods and services from those of competitors represents an additional support for *innovation*, enabling firms to capture *market share*, which contributes to *growth in our economy*. The granting and protection of *intellectual property rights* (IPRs) is vital to promoting *innovation* and *creativity* and is an *essential element of our free enterprise, market-based system*.

The U.S. Department of Commerce

IPRs are a set of exclusive rights, awarded by a state, for a limited period of time, to inventors, artists, and authors in exchange for public disclosure of their inventions and creative works. The logic is simple and straightforward. IPRs create, nurture, and promote an innovation ecosystem and provide a secure environment to inventors, artists, and authors. It is thus believed that inventors, artists, and authors will be incentivized under these conditions to invent/create original works as they can reap the benefits of their innovations. However, innovations impact not only the originators, but also the users in a remarkable manner. Thus, the invention of the electric bulb (U.S. Patent #223898; January 27, 1880) may have helped Thomas Edison to become rich and famous, but it has ultimately revolutionized the quality of life for everyone in the world ever since. Similarly, the creation of Apple personal computer (U.S. Patent #268584; April 12, 1983) may have helped Steve Jobs and Steve Wozniak, but this paved the way for many more innovations from Apple—such as iPod, iTunes, iPad, MacBook Air, iPhone, Apple Watch, etc.—which undoubtedly changed the way we live and communicate forever. Thus, innovations *benefit both the creators* and *users, create employment, increase trade* and *commerce, expand the economy,* and *improve the quality of life for everyone*. IPRs are *territorial* (except copyrights). Thus, every country must decide for itself how to structure its IPR regime based on its own vision and international norms. Not surprisingly, therefore, we find that many organizations and nations today are committed to *innovation* and IPRs.

Differences in the IPR Systems of the U.S., Europe, China, and India

As IPRs are territorial, differences exist in the IPR systems of different nations/regions. Since it is not possible to describe the IPR regimes of all the nations/regions in the world, we chose to outline a select set of IPR systems: *United States of America* (USA), *Europe, China,* and *India.*

The Intellectual Property Rights System of the USA

In the USA, there are three forms of IPRs namely—*patents, trademarks,* and *copyrights.* The first U.S. Patent Act was drafted into the U.S. Constitution in 1790. Accordingly, the *first* U.S. *patent,* numbered X000001, was granted by Thomas Jefferson to Samuel Hopkins for "making of pot ash and pearl ash by a new apparatus and process" in Pittsford, Vermont, on July 31, 1790. Thus, the history of the intellectual property system in the USA dates back to more than 225 years. Further, the *first numbered* U.S. *patent* was granted in 1836.[1a] The *patent* laws underwent a general revision, which was enacted July 19, 1952 and came into effect January 1, 1953. The U.S. *patent* law had been codified in **Title 35, United States Code (U.S.C.)**. Additionally, on November 29, 1999, the U.S. Congress enacted the **American Inventors Protection Act (AIPA)** of 1999, which further revised the *patent* laws. See Public Law 106-113, 113 Stat. 1501 (1999). Indeed, the USA is one of the most innovative nations in the history of the world. Thus, while it took 75 years for the USA to get to the 1,000,000th *patent,* it only took 6 years to go from 7 to 8 million *patents,* showing the acceleration of the pace of innovation.

In the USA, two agencies are central to the IP system:

1. The *United States Patent and Trademark Office* (USPTO)[1b] and
2. The *United States Copyright Office* (USCO).[2a]

The *United States Patent and Trademark Office*[1c] grants *patents* and registers *trademarks.* The USPTO also *counsels* the U.S. President, secretary of commerce, and other agencies of the U.S. government on IP *policy,* IP *protection,* and IP *enforcement;* further, the USPTO strives for effective IP protection of the U.S. *business interests* worldwide, through *free trade* and *international agreements.* In addition, the USPTO *provides training, education,* and *capacity building programs designed to foster respect for* IP and *encourage the development of strong* IP *enforcement regimes by the* U.S. *trading partners.*

The USCO and the position of *Register of Copyrights* were created in 1897.[2a] The USCO *manages* complex laws that include—registration, recordation of title and licenses, statutory licensing provisions, and other aspects of the *1976 Copyright Act*[2b] and the *1998 Digital Millennium Copyright Act.*[2c] It is important to note that the *Register* directs the *U.S. Copyright Office* as a separate department *within the Library of Congress.* Further, by statute, the *Register of Copyrights*: (a) serves as the principal advisor to the U.S. Congress on national and international *copyright* matters, (b) provides unbiased leadership and expertise to the U.S. Congress on *copyright* law and policy, and (c) testifies on request. Thus, the U.S. Congress relies upon, and directs, the USCO to provide critical law and policy services, including domestic and international policy analysis, legislative support for Congress, litigation support, assistance to courts and executive branch agencies, participation on U.S. delegations to international meetings, and public information and education programs.

Besides *patents, trademarks* (TMs), and *copyrights,* there is also the fourth form of IP in the USA, namely, the *trade secrets.* A *trade secret* is *confidential information* that a company uses in its own business, which has *independent economic value,* which only a *select few in the firm know,* and which the *company scrupulously protects from all the others both inside* and *outside the firm.* The USA has been a member of the *World Trade Organization* (WTO)[3a] since 1995. As per membership rules, all WTO member nations must give IPR protection to one another. As a result, the IP laws and enforcement procedures of different member nations are *closer rather than very different from each other,* as commerce and trade take priority. Further, the USA has been a party to the *Agreement on Trade-Related Aspects of Intellectual Property Rights* (TRIPS).[3b]

According to the USPTO:

> The USA fulfills its obligation by offering *trade secret protection* under the *state laws.* While *state laws* differ, there is similarity among the laws because almost all states have adopted some form of the *Uniform Trade Secrets Act* (see TRIPS).[3c]

In addition, the USA has been a signatory to the following IPR agreements:

- **The 1883 Paris Convention** for the Protection of Industrial Property[4a]—the USA had signed on to this international convention in 1913. This convention applies to the industrial property in the broadest sense, including *patents, trademarks, industrial designs, utility models* (a kind of "small-scale patent" provided for by the laws of some countries), *service marks, trade names* (under which an industrial or commercial activity is carried out), *geographical indications*

(indications of source and appellations of origin), and *the repression of unfair competition.*

- **The 1886 Berne Convention** for the Protection of Literary and Artistic Works[4b]—the USA had become a signatory to this convention in 1989. Under this convention, every member state fully recognizes the *copyrights* of authors in other member states on par with the *copyrights* of its own citizens.
- **The 1891 Madrid Protocol** for the International Registration of Trademarks[4c]—the USA signed onto the Madrid Protocol in 2003, which makes international registration of TMs possible. It is a cost-efficient way for individuals and businesses to apply for TM protection in multiple countries, through *a single application, a single office, using one language, one set of fees,* and *in one currency.*
- **The 1970 Patent Cooperation Treaty** for international *patent* application[4d]—In 1970, the USA became signatory to the Patent Cooperation Treaty (PCT). According to the *World Intellectual Property Organization* (WIPO)[5]:

> The PCT makes it *possible to seek patent protection for an invention simultaneously in each of a large number of countries* by filing an *international patent application.* Such an application may be filed by anyone who is a national or resident of a PCT Contracting State. It may generally be filed with the national *patent* office of the contracting state of which the applicant is a national or resident or, at the applicant's option, with the International Bureau of WIPO in Geneva.

Finally, it must be noted that the USA is not a signatory to **The Hague Agreement** for the international registration of *industrial designs,*[6] which makes the registration of *industrial designs* possible in multiple countries, through a single filing.

The Intellectual Property Rights System of the European Union

In the EU, there are five forms of IPRs: *patents, copyrights, trademarks, industrial designs,* and *geographical indications.*[7a] Unlike in the USA, multiple organizations are central to the IP system in the EU. Some of them are listed below:

1. *World Intellectual Property Office* (for all IP laws, treaties, and conventions);
2. *European Patent Organization*[7b] (for *patents*);
3. *European Copyright Office*[7c] (for *copyrights*);
4. *European Union Intellectual Property Office*[7d] (for *trademarks* & *designs*);

5. *European Commission* (EC)[7e] (for *geographical indications*, GIs); and
6. *World Trade Organization*[7f] (for global rules of trade).

According to the European Commission (EC):

> The *Single Market Strategy*[8a] is the European Commission's plan to unlock the full potential of the *Single Market*. The *Single Market* is at the heart of the European project, enabling people, services, goods, and capital to move more freely, offering opportunities for European businesses and greater choice and lower prices for consumers. It enables citizens to travel, live, work, or study wherever they wish.

Indeed, the goal of the EU is to provide IPR protection across the union.[8b] Thus, while the IPRs of a country in EU are protected by the national laws of that individual country, the *intellectual property laws* themselves are guided by the WIPO. The individual EU organizations mentioned above deal with the specific forms of IP as shown.

Intellectual Property Rights System in China

In China,[9] there are five specific forms of IPRs—*patents*, *copyrights*, *trademarks*, *designs*, and *geographical indications*. For many historical reasons, *the intellectual property rights system in China* began only recently, in the late 1970s. However, China fast-tracked setting up its IPR system soon after starting its economic reforms and began to open its economy to the rest of the world. China's goals were and still continue to be: (a) stimulate social progress, (b) meet the needs of developing a socialist market economy, and (c) accelerate China's entry into the global economy.[10] According to the Chinese government, an IPR system is critical to advance science and technology, develop the economy, and enrich culture.

Thus, China became a member state of the WIPO in 1980. Next, the *Trademark Law* of the People's Republic of China became effective on March 1, 1983, marking the beginning of China's modern legal IPR system. In continuation, while China became a signatory of the Paris Convention for the Protection of Industrial Property on March 19, 1985, the *Patent Law* of the People's Republic of China, came into effect on April 1, 1985.

According to the Information Office, State Council of the People's Republic of China,

> The *General Principles of the Civil Law of the People's Republic of China* were adopted at the fourth session of the Sixth National People's Congress on April 12, 1986, effective January 1, 1987. In this legislation, *intellectual property rights* as a whole were clearly defined in China's basic civil law, for the first time, *as the civil rights of citizens and legal persons*. This law, for the first time, *affirmed the citizens' and legal persons' right of authorship* (copyright law).[11]

In 1989, China became a signatory to the *Washington Treaty* in respect of *layout designs* (topographies) *of integrated circuits*.[12] Further, in the same year, China also became a member state of the *Madrid Agreement* for the International Registration of Trademarks. The *Copyright Law of the People's Republic of China* was adopted by the 15th meeting of the *Seventh National People's Congress Standing Committee* on September 7, 1990, effective June 1, 1991. In 1992, China became a signatory to the *Berne Convention* for the Protection of Literary and Artistic Works. In 1994, China became a signatory to the PCT of the WIPO. China joined the WTO and became its 143rd member in 2001 and accordingly developed a comprehensive system of IP laws that generally match international standards. Further, in 2007, China also became a member accepting the 2005 Protocol of the TRIPS.[13] Some organizations central to the IPR system of China are[14]:

1. *State Run Intellectual Property Office*[14a] (*patents*, integrated circuits);
2. *China Trade Mark Office*[14b] (*trademarks, geographical indications*); and
3. *Copyright Protection Center of China*[14c] (*copyrights*).

Intellectual Property Rights System in India

In India, there are nine specific forms of IPRs, namely—*patents, copyrights, trademarks, trade secrets, industrial designs, integrated circuits, plant varieties, GIS,* and *traditional knowledge*. India's Intellectual Property Rights system[15] has its beginnings in the colonial era.

Thus, India's IPR system can be recognized under two time-periods: *colonial* and *post-colonial*. In the *colonial* era, the first legislation in India relating to patents was the *Patent Act VI* of 1856. The goals of this legislation were to promote inventions and encourage inventors to disclose their inventions in exchange for some exclusive rights. This Act was replaced by the *Patent Act IX* of 1857. The Indian IPR system further evolved with few more legislations during 1857—1911. However, the 1911 *Indian Patents and Designs Act* (1911 IPDA), replaced all of the previous Acts, and brought the *patent* administration under the management of the Controller of Patents. 1911 IPDA was further amended in 1920 to secure priority with the United Kingdom and other countries.

According to the Controller General of Patents, Designs, and Trademarks of India:

> In 1930, further amendments were made to incorporate, inter-alia, provisions relating to grant of secret *patents, patent* of addition, use of invention by government, powers of the controller to rectify register of *patent*, and increase of term of the *patent* from 14 years to 16 years. In 1945, an amendment was made to provide for filing of provisional specification and submission of complete specification within nine months.[15]

The most significant development in the Indian IPR system in the *post-colonial* era is the introduction of the 1970 *Patent Act of India,* which repealed and replaced the 1911 IPDA and its *patents* laws. However, the 1911 IPDA continued its relevance for designs. The 1970 *Patent Act of India* came into force in 1972 with publication of the *Patent Rules.* It is important to point out how the Indian IPR system strategically evolved in the *post-colonial* era, taking into consideration the needs of the Indian pharmaceutical industry.[16] As per the 1970 *Patent Act of India,* the life of a *patent* was 14 years from the date of filing. Exceptions to this included "process patents" for preparation of drugs and food items whose term was only 7 years from the date of the filing or 5 years from the date of the *patent* grant, whichever is earlier. Further, Indian *patent* law allowed no "product patents" for drugs and food items, with the aim of building its own pharmaceutical industry. This empowered small Indian generic drug manufacturing companies to adopt and adapt the inventions of major international pharmaceutical companies at minimal Research & Development (R&D) costs. In effect, this policy of the Government of India contributed to an incredible growth in the licensed drug manufacturers from 2,237 in 1970 to 16,000 in 1993. Further, according to TechDirt[16]:

> The production of drugs grew at an average rate of 14.4% per year from 1980 to 1993, India became a *net exporter of pharmaceutical products*, and the market share of foreign multinational corporations dropped from 80%–90% to 40%. In 1995, six of the top ten pharmaceutical firms in India were domestic, and employment in the sector had reached half a million people.

Since then, these rules were periodically amended, while in 2003, *New Patents Rules* were introduced to replace the 1972 rules. These rules were amended in 2005 and 2006. The *Patents (Amendment) Act* 2006 with its amended *patent* rules has become effective from May 5, 2006. India has been a signatory to the *General Agreement on Trade and Tariffs* since July 8, 1948. India also became a WTO member on January 1, 1995 and a signatory to the TRIPS agreement since then. Furthermore, India accepted the 2005 *Protocol of the TRIPS Agreement* on March 26, 2007 to comply with international norms. This resulted in the Indian *Patent* system *accepting product patents in pharmaceutical innovations, extending the terms of all patents from* 5–14 years to 20 years, and *accepting the limitations on compulsory licensing.* The organizations central to the IPR system of India are[17]:

1. *Indian Patents Office*[17a] *(patents);*
2. *Copyright Office Government of India*[17b] *(copyrights);*
3. *Semiconductor Integrated Circuits Layout-Designs Registry*[17c] *(integrated circuits layout designs);*
4. *Protection of Plant Varieties & Farmer's Rights Authority of India*[17d] (plant varieties);
5. *National Biodiversity Authority*[17e] (traditional knowledge & biodiversity);

6. *Geographical Indications Registry*[17f] (*geographical indications*); and
7. *National Intellectual Property Organization*[17g] (IPR policies & management)

International Treaties and Conventions

There are 26 *international treaties* administered by WIPO,[18a,b] classified into three categories:

The IP Protection Treaties

This is a general group of 15 *international treaties* that *define mutually agreed basic standards of IP protection* by the member states:

1. Beijing Treaty on Audiovisual Performances;
2. Berne Convention;
3. Brussels Convention;
4. Madrid Agreement;
5. Marrakesh VIP Treaty;
6. Nairobi Treaty;
7. Paris Convention;
8. Patent Law Treaty;
9. Phonograms Convention;
10. Rome Convention;
11. Singapore Treaty on the Law of Trademarks;
12. Trademark Law Treaty;
13. Washington Treaty;
14. WIPO Copyright Treaty; and
15. WIPO Performances and Phonograms Treaty

Global IP Protection System Treaties

These are six *international treaties* that ensure that *one international registration or filing provides global protection for a given IP right* in all of the member states. Under these treaties, the services by WIPO simplify the procedure and reduce the filing costs of applications:

1. Budapest Treaty;
2. Hague Agreement;

3. Lisbon Agreement;
4. Madrid Agreement;
5. Madrid Protocol; and
6. The PCT

The International IP Classification Agreements

These are the four agreements that *allow the information on inventions, trademarks, and industrial designs to be indexed and organized into easily retrievable and manageable structures.*

1. Locarno Agreement;
2. Nice Agreement;
3. Strasbourg Agreement; and
4. Vienna Agreement

Other Agreements

These are the three other agreements, listed below:

1. Agreement between United Nations and WIPO;
2. Agreement between WTO and WIPO; and
3. Agreement on Trade-Related Aspects of Intellectual Property Rights (TRIPS)

WIPO Lex[19]

According to the WIPO:

> WIPO Lex is a *global database* that provides free access to some 12,000 national laws and treaties on IP from some 200 countries which are WIPO, WTO, or United Nations members. The WIPO Lex project was made possible through a common endeavor of the member states and other relevant bodies who contribute continuously to enriching the collection. WIPO invites member states and other IP stakeholders to expand and update the content by sending inputs and suggestions through WIPO-WTO common portal (IP authorities only) or through contact page (open to all).

The following is a detailed description of some of the most important *international treaties* agreements on *patents, copyrights,* and *designs.*

The Paris Convention

The "Paris Convention"[20a] for the protection of industrial property was adopted in 1883. It applies to many forms of industrial property including *patents, trademarks, service marks, industrial designs, utility models (small scale patents* granted by some countries), *trade names* (*designations* under which industrial/commercial activity is carried out), *geographical indications,* and *the repression of unfair competition.* The Paris Convention is the first international agreement between many nations to provide protection to the intellectual works of one state among all other member states. The provisions of the convention are classified into three categories: *National Treatment, Right of Priority,* and *Common Rules.*[20b]

National Treatment

According to these provisions, each contracting state must grant the same protection of industrial property to the nationals of other contracting states that it grants to its own nationals. Further, even the nationals of non-contracting states are entitled to *National Treatment* if they are domiciled or have a real and effective industrial or commercial establishment in a contracting state.

The Right of Priority

Under these provisions, an application for IP protection—*patents* (or *utility models* where applicable), *marks,* and *industrial designs*—in one member state receives the *Right of Priority* in all of the contracting states. In other words, an applicant who made the first application for IP protection in one of the contracting states is granted sufficient time (12 months for *patents/utility models* and 6 months for *industrial designs* and *marks*) to subsequently apply for IP protection in other contracting states of choice. What is important to note is that all subsequent applications by this applicant in the other contracting states will be considered as if they had been filed on the same day of the first IP application.

Two corollaries result from this:

1. the first application filed by an applicant in a contracting state will have priority (and hence the *Right of Priority*) over all other applications by the others for the same IP (invention, *utility model, mark,* or *industrial design*) during the said period of time and
2. subsequent applications will be unaffected by any event in the interim—such as the publication of an invention or incorporating an *industrial design* or the sale of articles bearing a *mark*—as they are all based on the first application.

According to the WIPO[20b]:

> One of the great practical advantages of this provision is that applicants seeking protection in several countries are not required to present all of their applications at the same time, but have 6 or 12 months to decide in which countries they wish to seek protection, and to organize with due care the steps necessary for securing protection.

Common Rules

Under these provisions, the Paris Convention outlines some *Common Rules* regarding *patents, marks, industrial designs, indications of source, and unfair competition,* which all contracting states must follow. Some of those important rules are listed below:

1. ***Patents.*** The *patent* granting process stays independent among the member states for the same invention. Thus, no contracting state is obliged to grant a *patent* for a particular invention because another member state granted a *patent* for it. On the contrary, no contracting state can refuse, annul, or terminate a *patent* for a given invention because another member state had done so. Similarly, a *patent* may not be refused or invalidated by a contracting state because the production or sale of the resulting product is subject to restrictions or limitations due to domestic laws. A contracting state may *provide compulsory licenses to patents to prevent abuses by the monopolistic rights of a patent under certain conditions,* only as outlined by the *Common Rules* of the Paris Convention.
2. ***Marks.*** The filing and registration of *marks* are *not regulated* by the Paris Convention and are determined by the contracting states based on their own domestic laws. Thus, the registration of a *mark* in one contracting state is independent of its registration in any other contracting state, including the country of origin. Consequently, no one's application for registration of a *mark* can be denied or a registration invalidated in a contracting state because it had not been filed, registered, or renewed in the country of origin. When a *mark* had been duly registered in the country of origin, and a filing made for the same in any other contracting state, it must be accepted and protected. However, a *filing can be denied* by a contracting state *on any of the following justifiable grounds*: (a) when there is infringement on the acquired rights of third parties; (b) when it is not distinctive; (c) when it is immoral or offensive to public; and (d) when it is deceptive in nature. There are more rules in this regard.[18b]
3. ***Industrial Designs.*** A contracting state of the Paris Convention must protect and not forfeit the *industrial design* of another contracting

state on the grounds *that the article incorporating the design is not manufactured in that state.*

4. ***Indications of Source.*** All contracting states must take effective legislative and enforcement measures *to prevent/stop false indication of the source of goods or false identity of the manufacturer/trader.*
5. ***Unfair Competition.*** All contracting states must have effective legislative and enforcement mechanisms *against unfair competition.*

The Paris Convention which came into force in 1884, was revised many times—Brussels (1900), Washington (1911), Hague (1925), London (1934), Lisbon (1958), Stockholm (1967), and was amended in 1979. Today, there are 176 countries who are signatories to the Paris Convention.

The Berne Convention

Copyright protection at international levels began approximately 150 years ago in terms of bilateral treaties. Unfortunately, such bilateral treaties were neither comprehensive nor consistent. As a result, there was a real need for a system of *copyright protection* that was wide-ranging and uniform. This led to the formulation and adoption of the "Berne Convention"[21a] in 1886, whose goal was, *to protect, in as effective and uniform a manner as possible, the rights of authors in their literary and artistic works.*

The Berne Convention is based on *the three principles* and contains *the general provisions* that determine the minimum standards of *copyright protection* and *special provisions* that are available to developing countries[21b]:

The Three Principles

1. ***National Treatment Principle.*** The literary and artistic works that originate in one contracting state must be protected equally by all the other contracting states just as they would protect the works of their own nationals;
2. ***Automatic Protection Principle.*** *National Treatment* must not be conditional upon compliance with any formality; and
3. ***Principle of Independence of Protection.*** Protection of literary and artistic works in a contracting state is independent of the existence of protection in the country of origin of the works. However, when a contracting state provides longer term protection than the minimum prescribed by the Berne Convention, and the protection for the work ceases in the country of origin, then the contracting state may also deny protection soon after.

General Provisions

Under these provisions that determine minimum standards of *copyright protection*, a state must guarantee *copyright protection*[21b] that include: (a) works; (b) rights to be protected; and (c) the duration of protection.

1. *Works.* As per Article 2(1) of the convention, every work in the literary, scientific, and artistic domain must be protected regardless of its mode/form of expression.
2. *Rights to be Protected.* The Berne Convention authorizes the following *exclusive rights*, as the rights to be protected:
    a. *Translation Right*: The right to translate own work into other languages,
    b. *Adaptation Right*: The right for adaptations and arrangements of own work,
    c. *Public Performance Right*: The right to perform one's own work in the public in a drama, musical-drama, and musical works,
    d. *Recitation Right*: The right to recite one's own literary works in public,
    e. *Communication Right*: The right to communicate about the performance of one's own work to the public,
    f. *Broadcasting Right*: The right to broadcast one's own work (with the possibility that a contracting state may provide for a mere right to equitable remuneration instead of a right of authorization),
    g. *Reproduction Right*: The right to make reproductions by any method or in any form (with the possibility that a contracting state may permit, in special cases, reproduction without authorization, provided that the reproduction does not conflict with the normal exploitation of the work and does not unreasonably prejudice the legitimate interests of the author; and the possibility that a contracting state may provide, in the case of sound recordings of musical works, for a right to equitable remuneration), and
    h. *Right to Audiovisual Works*: The right to use one's own as the basis for an audiovisual work, and the right to reproduce, distribute, perform in public, or communicate to the public that audiovisual work.

    In addition, the Berne Convention authorizes certain "moral rights," such as—(a) the right to claim authorship of the original work, (b) the right to object to any mutilation/

deformation/modification of the work, and (c) the right to object to derogatory/prejudicial actions harmful to the author's honor or reputation.

3. *Duration of Protection*. The general rule is that *protection must be granted until the expiration of the 50th year after the author's death*. According to WIPO, there are exceptions to this general rule[21b]:

> In the case of *anonymous* or *pseudonymous works*, the term of protection ends 50 years after the work has been lawfully made available to the public, except if the pseudonym leaves no doubt as to the author's identity or if the author discloses his or her identity during that period; in the latter case, the general rule applies. In the case of *audiovisual (cinematographic) works*, the minimum term of protection is 50 years after the making available of the work to the public ("release") or—failing such an event—from the creation of the work. In the case of *works of applied art* and *photographic works*, the minimum term is 25 years from the creation of the work.

Special Provisions Available to Developing Countries

According to the Appendix to the *Paris Act of the Berne Convention*,[21b] *developing countries are allowed to implement non-voluntary licenses for translation and reproduction of works in certain cases, in connection with educational activities. In these cases, the described use is allowed without the authorization of the right holder, subject to the payment of remuneration to be fixed by the law.*

The Berne Convention which came into existence in 1886 was revised in 1896 (in Paris), 1908 (in Berlin), 1914 (in Berne), 1928 (in Rome), 1948 (in Brussels), 1967 (in Stockholm), 1971 (in Paris), and amended in 1979. Today, there are 172 contracting states to the Berne Convention. According to the Berne Convention, instruments of ratification or accession must be deposited with the Director General of WIPO.

The Patent Cooperation Treaty

PCT[22] is an international agreement in the field of *patents* administered by WIPO. Arguably, it is the most important advance in the area of international cooperation on patenting since the Paris Convention. PCT provides a systematic, rational, and cooperative mechanism for filing, searching, examining, and disseminating the technical information for an "international patent application." However, PCT does not provide for grant of any "world-wide patent," as such a thing does not exist. The responsibility of granting *patents* is still left exclusively to the *patent* offices of the various countries wherein protection is sought (the "designated offices"). PCT is complimentary to the Paris Convention and is not a competitor to it. Indeed, this special treaty is open to all the 140-member countries of the Paris Convention.

Thus, PCT makes simultaneous *patent* protection of an invention possible in a large number of countries through a single filing of an "international *patent* application." Such a *patent* application may be filed by anyone who is a national/resident of a PCT Contracting State, as per two filing options: (a) it may be filed with the national *patent* office of the contracting state of which the applicant is a national/resident, or (b) it may be filed with the International Bureau of WIPO in Geneva. The chief goal of the PCT is *to rationalize the previously established procedures for patent protection in several countries, to serve the interests of the users, as well as benefit the offices which have the responsibility for administering it*. PCT regulates the formal requirements of the "international patent applications," in two phases: (1) the *international phase* and (2) the *national phase,* as outlined below:

The International Phase

1. ***International Filing Date***. The filing date of a PCT application—filed by an applicant in a particular contracting state or with the International Bureau of WIPO in Geneva, in one language and paying one set of fees—becomes binding in the national *patent* offices of all contracting states. Thus, a PCT application automatically binds all contracting states to the *international filing date*. When there is no priority claim, the filing of an international application can be immediate (0 months). On the other hand, when there is a priority claim, the filing of an *international patent application* can be done 12 months from the priority date.
2. ***International Search Report and Written Opinion***. An international *patent* application requires an "international search," conducted by one of the designated and competent "international searching authorities" (ISA). The results of international search are captured in an *international search report,* which cites all published documents that might affect the patentability of the invention claimed in the international application. Further, a preliminary and non-binding *written opinion* by ISA also outlines its opinion on whether the invention meets the patentability criteria in light of the results. The *international search report* and *written opinion* are then communicated to the applicant. This process is expected to take 9 months from the international filing date (for applications without any priority claim) or 16 months from the priority date (for applications with a priority claim). At this stage, the applicant may opt to either: (a) amend the claims in the application or (b) withdraw the application itself, if the *patent* grant seems very unlikely.
3. ***International Publication***. When the international search does not find any invention-disqualifying results, then the *written opinion* is possibly favorable. Under these circumstances, the international

application (which is not withdrawn) and the *international search report* will be published by the international bureau. The purpose of the *international publication* is to disclose the content of the international application to the world, as well as provide provisional protection for the invention under consideration. This process is expected to take 18 months for applications from the international filing date or the priority date. However, at this time, the *written opinion* is not published.

4. ***Supplementary International Search (Optional).*** Within 19 months from the priority date, an applicant may exercise the option to request for *supplementary international search* from a "supplementary international searching authority" (an ISA willing to offer this service), in the particular language in which that authority specializes. The goal of this secondary search is to significantly minimize the likelihood of any additional art that could pose problems in the national phase.
5. ***International Preliminary Examination (Optional).*** At this stage, an applicant may choose to make changes to the application based on the results of the search reports and/or conclusions of the written opinion and submit an amended application to an "International Preliminary Examining Authority" for *international preliminary examination*. Thus, one of the ISAs (International Preliminary Examining Authority) carries out additional patentability analysis on a revised version of the application. The outcome of the international preliminary examination is an "International Preliminary Report on Patentability" (chapter II), which summarizes a preliminary nonbinding opinion on the patentability of the claimed invention. The applicant will now have an even more objective basis to appraise his/her chances of obtaining a *patent* and decide to advance the application to the national and regional *patent* offices or withdraw it altogether. In case no international preliminary examination has been requested, the international bureau establishes an "International Preliminary Report on Patentability" (chapter I) based on the written opinion of the ISA and forwards it to the designated offices.

The National Phase

At the end of the international phase (usually at 30 months from the earliest filing date of the initial application, from which the applicant claims priority), the applicant may pursue the grant of the *patents* directly before the designated national (or regional) *patent* offices of the countries of interest. This is the beginning of the *national phase*. For this, one needs to do the following: provide a translated version of the application in the official language of the individual office (where necessary), pay the required fees, and acquire the services of local *patent* agents.

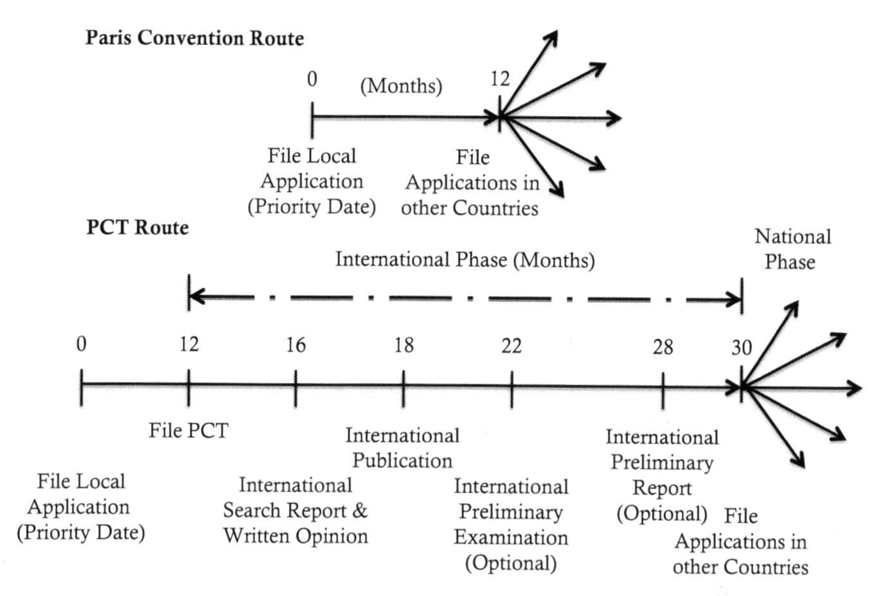

FIGURE 2.1
Comparison of the Paris Convention route and the PCT route.

Figure 2.1 schematically illustrates the key differences in patenting between the PCT route and the Paris Convention route.[21]

Differences between the PCT and the Paris Convention Routes. In the Paris Convention route, an applicant may choose to directly file separate, but simultaneous, *patent* applications in all the countries of interest (for some countries, regional *patents* may be available) or file first in a Paris Convention country, claim a priority date, and then file separate *patent* applications in other Paris Convention countries within 12 months from the filing date of that first *patent* application and claim the same priority date in all the other countries.

In the PCT route, an applicant may directly file an international *patent* application in a PCT contracting country or with the International Bureau of WIPO in Geneva. Alternatively, applicant can file a local *patent* application, claim priority, and then file an international *patent* application within the 12-month period from the filing date of first application, valid in all the contracting states of the PCT. Thus, the PCT route offers the following advantages: (a) bring all the countries (and markets) of patenting interest within reach; (b) postpone the major costs associated with *patent* protection for 30 months; and (c) provide stronger *patent* analysis for deciding whether or not to *patent* an invention in many countries. As a result, the PCT route is used by the world's major corporations, research institutions, and universities for international *patent* protection.

The PCT was amended in 1979 and modified in 1984 and in 2001. Instruments of ratification or accession must be deposited with the Director General of WIPO.

The Madrid System for International Trademark Registration

The "Madrid System"[23a] is a simple, convenient, and organized system for registering and managing *marks* worldwide. Thus, an applicant can file one application, in one language, and pay one set of fees to protect a trademark/service *mark* in the territories of up to 98-member states.

The Madrid System manages the portfolio of *marks* through one centralized system. The international trademark/service *mark* registration is achieved in three stages in the Madrid System[23b]:

Stage 1—Application through the National/Regional IP Office

Prior to filing an international application, an applicant must ensure that the mark is either registered or an application for it has already been filed in the "home" IP office (office of origin). That is known as the **basic *mark***. The applicant then needs to submit an international application for it through the same IP office, which certifies and forwards it to WIPO.

Stage 2—Formal Examination by WIPO

Once the international application is received, WIPO conducts a formal examination and approves it. Subsequently, the *mark* is recorded in the international register and published in the WIPO Gazette of International Marks. Then, WIPO sends the applicant *the certificate of international registration* and notifies the IP offices in all the territories in which the applicant seeks protection for the *mark*.

Stage 3—Substantive Examination by National or Regional IP Offices (Office of the Designated Contracting Party)

The IP offices of the territories where applicant wants to protect the *mark* will make a decision within the applicable time limit (12 or 18 months) as per their legislation. WIPO will record the decisions of the IP offices in the International Register and then notify the applicant.

The international registration of a *mark* is valid for 10 years. It is possible to renew the international *mark* at the end of each 10-year period directly with WIPO with effect in the designated contracting parties.

The Hague Agreement for the International Registration of Industrial Designs

The "Hague Agreement"[24a] for *the International Registration of Industrial Designs* makes the registration of up to 100 designs possible, in more than 66 territories, via a single international filing.

The Hague Agreement, which concluded in 1925, was revised in 1934 (London) and 1960 (Hague), completed by an additional Act in 1961 (Monaco), complementary Act in 1967 (Stockholm), amended in 1979, and adopted in 1999 (Geneva). Today, two "Acts of The Hague Agreement" are in operation—the 1999 Act and the 1960 Act.[24b] In September 2009, the application of the 1934 Act of The Hague Agreement was frozen to simplify and streamline the international design registration system.

The Hague Agreement[24c] allows applicants to file a single international application with the International Bureau of WIPO to protect their designs in multiple countries or regions in a simple manner. Further, The Hague Agreement also requires a simple procedure for subsequent management of an *industrial design* registration—such as, recording of subsequent changes and renewal of the international registration—through a single step. It is important to note that an international application may be governed by the 1999 Act, the 1960 Act, or both, depending on the contracting party. International design applications may be filed with the International Bureau of WIPO, either directly or through the industrial property office of the contracting party of origin, if the laws of that contracting party require so. In practice, however, nearly all international applications are filed directly with the International Bureau of WIPO, using its electronic filing interface. The 1960 Act is open to all the member states of the Paris Convention (1883), while the 1999 Act encourages more prospective contracting parties to join.

The TRIPS Agreement

The WTO[25a] is an international organization that deals with the rules of trade among nations. The TRIPS Agreement of WTO[25b]—negotiated in the 1986–1994 Uruguay Round—introduced, for the first time, the rules for IPRs into the multilateral trading system between nations. By becoming members of the WTO, countries agree to comply with 18 specific agreements annexed to the Agreement establishing the WTO. Thus, WTO members cannot choose to be party to some agreements and not the others (except a few non-obligatory agreements).

The TRIPS Agreement aims to narrow the gaps in protection given to the IPRs around the world, by bringing them under the purview of common international rules. TRIPS institutes the minimum levels of protection that each government has to guarantee to the intellectual property of fellow WTO members—striking a balance between *short-term costs* and *long-term benefits* to the society. As outlined in Chapter 1, society benefits in the long term, as IPRs have been proven to encourage creativity and innovation, especially after the expiration of protection and when the creations and inventions become a part of the public domain. Thus, WTO's dispute settlement system is now available to the trading nations when disputes arise over IPRs.

The TRIPS Agreement covers the following five broad topics[25c]:

- *Basic principles* of the trading system and *key international agreements* on IPRs;
- *Ground rules* for *providing adequate protection* to IPRs;
- *Mechanisms* for *enforcement* of the IPRs in their territories;
- *Methods* for the *settlement of disputes* on IPRs between WTO member states; and
- *Transitional arrangements,* when the new system is being introduced.

The areas of IP covered[25c] under the TRIPS Agreement are: *patents, copyrights, trademarks/service marks, layout-designs of integrated circuits, industrial designs, geographical indications,* and *undisclosed information, including trade secrets.*

Basic Principles

The TRIPS Agreement mandates three basic principles for IPR protection: (a) *National Treatment,* according to which a member nation should treat one's own nationals and foreigners equally; (b) *most-favored nation treatment,* according to which a member nation should provide equal treatment for nationals of all the trading partners in the WTO; and (c) *enhancement of economic* and *social welfare,* according to which *IPR* protection among the member nations should contribute to innovation and technology transfer, such that the producers and users of IP benefit.

Common Ground Rules for Adequate IPR Protection

The TRIPS Agreement sets the common ground rules in terms of the international conventions for different kinds of IPRs that the member states must follow: (a) the Paris Convention for the Protection of Industrial Property (*patents, trademarks/service marks, industrial designs, utility models, trade names, geographical indications,* and *the repression of unfair competition*) and (b) The Berne Convention for the Protection of Literary and Artistic Works (*copyrights*). For details, see WTO website.[25d]

Mechanisms for Enforcement of the IPRs in the Territories

The TRIPS Agreement emphasizes the enforcement of IPRs. According to the TRIPS Agreement[25b]:

> Governments have to ensure that intellectual property rights can be enforced under their laws and that the penalties for infringement are tough enough to deter further violations. The procedures must be fair and equitable and not unnecessarily complicated or costly. They should not entail unreasonable time limits or unwarranted delays. People involved should be able to ask a court to review an administrative decision or to appeal a lower court's ruling.

As discussed above, the TRIPS Agreement elaborates how enforcement should be handled, including rules for obtaining evidence, provisional measures, injunctions, damages, and other penalties. TRIPS demands its member states to treat intentional *trademark* counterfeiting or *copyright* piracy on a commercial scale as a criminal offence. Accordingly, courts should exercise their authority, under certain conditions, to order the disposal/destruction of pirated/counterfeit goods. Further, it asks the governments to ensure the IPR owners the help and assistance they need from the customs authorities to prevent imports of counterfeit and pirated goods.

Methods for Settlement of Disputes

WTO also outlines methods for the settlement of disputes on IPRs between the member nations. These can be found in the WIPO Intellectual Property Handbook on Policy, Law, and Use.[26]

Transitional Arrangements

WTO's Agreements came into effect on January 1, 1995. WTO granted the following transitional deadlines to its member nations in order to ensure that their laws and practices confirmed with the TRIPS Agreement:

1. Developed countries: 1 year (until January 1, 1996);
2. Developing countries: 5 years (until January 1, 2000); and
3. Least-developed countries: 11 years (until January 1, 2011, now extended to 2013 in general; until 2016 for pharmaceutical *patents* and undisclosed information)

Subject to a few provisions, the guiding principle is that the member nations of the TRIPS Agreement comply with their obligations to the IPRs at the end of the country's transition period, as well as the new ones. The details of all the other treaties and agreements can be found elsewhere.[26]

References

1. (a) https://www.uspto.gov/web/offices/ac/ido/oeip/taf/issudate.pdf; (b) https://www.uspto.gov/; and (c) USPTO: The name of the US Patent Office was changed first to the "Patent and Trademark Office" (PTO) in 1975 and then to the "United States Patent and Trademark Office" (USPTO) in 2000.
2. (a) USCO: http://www.copyright.gov/; (b) https://www.copyright.gov/title17/title17.pdf; (c) https://www.copyright.gov/legislation/pl105-304.pdf
3. (a) WTO: https://www.wto.org/; (b) TRIPS: http://www.uniformlaws.org/shared/docs/trade%20secrets/utsa_final_85.pdf; (c) https://www.uspto.gov/patents-getting-started/international-protection/trade-secret-policy
4. (a) Paris Convention: http://www.wipo.int/treaties/en/ip/paris/; (b) Berne Convention: http://www.wipo.int/treaties/en/ip/berne/; (c) Madrid Protocol: http://www.wipo.int/treaties/en/registration/madrid_protocol/; and (d) Patent Cooperation Treaty: http://www.wipo.int/treaties/en/registration/pct/
5. WIPO: http://www.wipo.int/treaties/en/registration/pct/summary_pct.html
6. The Hague Agreement: http://www.wipo.int/edocs/pubdocs/en/designs/911/wipo_pub_911.pdf
7. (a) Europa: http://europa.eu/youreurope/business/start-grow/intellectual-property-rights/index_en.htm; (b) EPOrg: https://www.epo.org/about-us/organisation.html. The *European Patent Organization*, has two bodies: The *European Patent Office* (EPO) and the *Administrative Council*. EPO offers inventors a uniform *patent application procedure* which enables them to seek patent protection in up to 42 countries. EPO is supervised by the *Administrative Council*, which is composed of representatives of the Organization's member states; (c) ECO: https://www.eucopyright.com/; (d) EUIPO: https://euipo.europa.eu/ohimportal/en; (e) EC: http://ec.europa.eu/index_en.htm;http://www.wipo.int/edocs/mdocs/geoind/en/wipo_geo_sof_09/wipo_geo_sof_09_www_124276.pdf; and (f) WTO: https://www.wto.org/index.htm
8. (a) http://ec.europa.eu/growth/single-market/ and (b) http://ec.europa.eu/growth/tools-databases/newsroom/cf/itemdetail.cfm?item_id=8580&lang=en&title=Have-your-say-on-the-enforcement-of-intellectual-property-rights
9. https://www.gov.uk/government/uploads/system/uploads/attachment_data/file/456352/IP_rights_in_China.pdf
10. http://www.china-un.ch/eng/bjzl/t176937.htm
11. China Government White Papers: http://www.china.org.cn/e-white/intellectual/12-2.htm
12. http://www.china.org.cn/e-white/intellectual/12-2.htm
13. TRIPS: https://www.wto.org/english/tratop_e/trips_e/amendment_e.htm
14. (a) SIPO: http://english.sipo.gov.cn/news/official/201611/t20161123_1302924.html; (b) CTMO: http://www.ctmo.gov.cn/; and (c) CPCC: http://www.ccopyright.com.cn/
15. History of the Indian Patent System: http://www.ipindia.nic.in/history-of-indian-patent-system.htm

16. Beginnings of the Generic Indian Pharmaceutical Industry: https://www.techdirt.com/articles/20090530/1620345062.shtml
17. (a) DIPP: http://dipp.nic.in/English/default.aspx; (b) COGoI: http://copyright.gov.in/; (c) SICLDR: http://sicldr.gov.in/; (d) PPV & FRA: http://plantauthority.gov.in/; (e) GIR: http://www.ipindia.nic.in/gi.htm; (f) NBA: http://www.nbaindia.org/; and (g) NIPO: http://www.nipo.in/index.htm
18. (a) WIPO Treaties: http://www.wipo.int/treaties/en/; and (b) http://www.wipo.int/export/sites/www/about-ip/en/iprm/pdf/ch5.pdf
19. WIPO Lex: http://www.wipo.int/wipolex/en/index.jsp?tab=1
20. Paris Convention: (a) http://www.wipo.int/treaties/en/ip/paris/; and (b) http://www.wipo.int/treaties/en/ip/paris/summary_paris.html/
21. Berne Convention: (a) http://www.wipo.int/treaties/en/ip/berne/; and (b) http://www.wipo.int/treaties/en/ip/berne/summary_berne.html
22. (a) http://www.wipo.int/pct/en/faqs/faqs.html; and (b) http://www.wipo.int/treaties/en/registration/pct/
23. (a) http://www.wipo.int/madrid/en/; and (b) http://www.wipo.int/madrid/en/how_madrid_works.html
24. (a) http://www.wipo.int/treaties/en/registration/hague/; (b) http://www.wipo.int/about-ip/en/iprm/; and (c) http://www.wipo.int/treaties/en/registration/hague/summary_hague.html
25. (a) https://www.wto.org/index.htm; (b) https://www.wto.org/english/thewto_e/whatis_e/tif_e/agrm7_e.htm; (c) For a detailed treatment of each of these subjects, see 25 (b); and (d) https://www.wto.org/english/tratop_e/trips_e/intel2_e.htm
26. WIPO. 2008. Arbitration and mediation of IP disputes, in *WIPO Intellectual Property Handbook on Policy, Law and Use*, pp. 230–234. WIPO Publication 489E; ISBN 978-92-805-1291-5.

Section II

The Full Landscape of Intellectual Property Rights

3

Patents

Patents provide *incentives* to individuals by recognizing their *creativity* and offering the possibility of *material reward* for their marketable *inventions.* These incentives encourage *innovation,* which in turn *enhances the quality of human life.*

World Intellectual Property Organization

Like the physical or financial property rights, *intellectual property rights* (IPRs) enable the creators/owners of IP to benefit from their investment in creative work. Indeed, Article 27 of the Universal Declaration of Human Rights[1] outlines these rights and awards the rights to benefit from the authorship of scientific, literary, or artistic productions. By IPRs, we mean *industrial property* (which include *patents, trademarks, industrial designs, geographical indications,* and *protection against unfair competition*) and *copyrights.* As discussed in chapter 2, the Paris Convention (1883) recognized the value of *industrial property protection* and the Berne Convention (1886) recognized the value of *copyright protection.*

Part-A: General Overview of Patents

Concept of a Patent

According to the World Intellectual Property Organization (WIPO)[2a]:

> A *patent* is "a document—issued upon application, by a *government office* (or a *regional office* acting for several countries)—which describes an *invention* and creates a legal situation in which the *patented invention* can normally only be exploited (manufactured, used, sold, imported) with the authorization of the owner of the *patent.*"

An *invention* is a solution to a specific problem and may relate to a product or a process.[2a]

In another source document, the WIPO states[2b]:

> A *patent* is an *exclusive right granted for an invention,* which is a product or a process that provides, in general, a new way of doing something or offers a new technical solution to a problem. To get a *patent, technical information about the invention must be disclosed* to the public in a *patent* protection.

Similarly, according to the United States Patent and Trademark Office (USPTO)[2c]:

> A *patent* for an invention is the grant of a property right to the inventor, issued by the United States Patent and Trademark Office.

According to the Government of India[2d]:

> A *patent* is granted for an invention which is a new product or process involving an inventive step and capable of industrial application.

Finally, in China:

> A *patent* is an exclusive right *granted for an invention, which is a product or a process that provides, in general, a new way of doing something or offers a new technical solution to a problem.* A *patent* provides protection for the invention to the owner of the *patent* and that protection means that the invention cannot be commercially made, used, distributed, or sold without the *patent* owner's consent. These *patent* rights are usually enforced in a court, which, in most systems, holds the authority to stop *patent* infringement.[2e]

Thus, we can note that *the concept of a patent* involves the following *key points*:

1. A *patent* is an *exclusive right;*
2. It is *given to an inventor;*
3. It is *given for an invention;*
4. The invention *must show a new/improved way to a product or process;*
5. The *technical information* that describes the invention *must be publicly disclosed;* and
6. A *patent* can be *issued only by a competent patent granting authority* (the national *patent* office of a country or a regional *patent* office of a region).[2f]

From the above discussion, it should be clear that a *patent* right is a *territorial right* (meaning *the right is effective only in the country which grants it*). For example, a United States (U.S.) *patent* is legally effective only in the United States, U.S. territories, and U.S. possessions. Also, an inventor who discloses his *invention will have no patent* until: (a) he files a *patent* application, (b) his work meets all statutory requirements, and (c) the *patent* granting authority awards him a *patent*.

Indeed, *patented inventions* are absolutely critical to every aspect of our life. Some examples are—the electric bulb (*patents* of Thomas Edison), plastics (*patents* of Baekeland), ballpoint pens (*patents* of Biro), and microprocessors (*patents* of Intel), to mention a few.[2g] Thus, it is important to recognize that a *patent* is *not just an abstract concept* and *involves the disclosure and protection of inventions* (products/processes) *potentially valuable to everyday living*. As a result, nations and governments reward inventors with *patents* and encourage them to innovate new or improved products/services to solve the needs, wants, and challenges of the world.

Nature of the Patent Right

According to the USPTO[2c]:

> The right conferred by the *patent* is—in the language of the statute and of the grant itself—*the right to exclude others* from making, using, offering for sale, or selling the invention in the United States or importing the invention into the United States. What is granted is *not the right to make, use, offer for sale, sell, or import*, but *the right to exclude others from making, using, offering for sale, selling, or importing the invention*. Once a *patent* is issued, the patentee must enforce the *patent* without the aid of the USPTO.

Thus, a *patent* right is *an exclusionary right* and *not an affirmative right*.

Purpose of Patent Protection

Individuals and companies seek *patents* to *protect* their inventions and innovations. Broadly speaking, individuals seek *patent* protection to *strengthen their offerings, improve their position for new venture creation*, and/or *license or sell their IP to create personal wealth*. On the contrary, companies seek *patent* protection *to fulfill much more complex* and *multidimensional business objectives* as indicated below:

1. *To deliver higher returns on investment to shareholders and investors*. The products/services of a company that are well-protected by *patents* make imitations by rival competitors very difficult. In addition, a *patent* can create a barrier for new entrants into the marketplace. As a result, such a company is more likely *to achieve sustainable revenues, profits*, and *market share*—which will in turn help it to deliver *higher returns on investment* to shareholders and investors.[3a]
2. *To achieve competitive advantage*. *Patents* with strong technical claims enable businesses to make *superior marketing claims* on their

products/services. Generally, as customers verify the marketing claims by their own use of the products/services, they will be willing to pay more for them. Thus, innovative companies, capable of making superior claims, could set higher prices for their products/services.

3. *To strengthen the ability to raise capital.* Companies with strong *patent* portfolios often earn "reputation and trust" in the marketplace as "innovative leaders." As a result, such companies can raise capital when needed—through low-interest bank loans and sale of lucrative company stock.
4. *To create an alternative revenue stream and/or test a new business idea.* Companies may choose to license/sell some of their *patents*—that are not either utilized in their current businesses (mission) or not useful in their future plans (vision)—to create an *alternative stream of revenue* and *avoid maintenance fees.* Alternatively, they may create a joint venture and test a new business idea at significantly lower risk.
5. *To achieve higher valuation of the firm.* Companies with strong portfolios of *patents* (the most tangible of all the intangible assets of a company) can skillfully communicate the same to investors so that it leads to *higher valuations of the firms.*
6. *To enforce patent rights.* Companies with strong *patent* portfolios *can enforce patent rights better* to prevent/stop counterfeit products/services through aggressive litigation.

NOTE: Starting from this point, this chapter predominantly uses information from the USPTO and WIPO sources for discussion and clarifications. Where relevant, however, essential information pertaining to India and China will be brought into the mix.

According to the United States Constitution (U.S.C.), Article I, Section 8,[3b]

> Congress shall have power... to promote the progress of science and useful arts, by securing for limited times to authors and inventors the exclusive right to their respective writings and discoveries.

Thus, the U.S. Constitution empowers the Congress to enact various laws relating to *patents.*

Accordingly, the first U.S. *patent* law was enacted in 1790. Subsequently, the U.S. *patent* laws were revised and codified in **Title 35, United States Code (35 U.S.C.).**[3c] This law was enacted on July 19, 1952 and came into effect January 1, 1953. Further, the U.S. Congress enacted the **American Inventors Protection Act of 1999**[3d] on November 29, 1999, which again revised the *patent* laws.

Thus, the U.S. *patent* law specifies the subject matter for which a *patent* may be obtained, clarifies the conditions for patentability, and empowers the United States Patent and Trademark Office to administer itself.

Patentable and Not-Patentable Subject Matter—A Brief Sketch

Patentable Subject Matter

According to 35 U.S.C. § 101[3e]:

> Whoever invents or discovers any *new* and *useful process, machine, manufacture,* or *composition of matter* or *any new and useful improvement thereof,* may obtain a *patent* therefor, subject to the conditions and requirements of this title.

The approximate definitions of these terms are:

- The term "process" is defined as *a process, act, or method, and primarily includes industrial or technical processes.*
- The term "machine" used in this statute is self-explanatory.
- The term "manufacture" refers to *articles that are made and includes all manufactured articles.*
- The term "composition of matter" relates to *chemical compositions and may include mixtures of ingredients as well as new chemical compounds.*

In essence, 35 U.S.C. § 101 provides for *two categories of patentable subject matter.* They are: (1) processes and (2) products (*machine, article of manufacture, composition of matter*). Further, this statute stipulates not only *new and useful products and processes,* but also *new and useful improvements of such products and processes.* Notably, 35 U.S.C. § 101 specifies that the subject matter must be *useful.* This term clarifies that the subject matter *must function to practically deliver a useful purpose.* These categories of inventions practically include ***anything under the sun that is made by a human*** **and** ***the processes required for making them.***

It is important to note that a *patent* is not given for *merely proposing an idea* or *making a suggestion for a product/process.* One can get a *patent only for an invention involving an actual, new/improved product* (machine, article of manufacture, or composition of matter) or *process,* for which the inventor provides complete description.

Not-Patentable Subject Matter

In continuation, the interpretations of 35 U.S.C. § 101 by the U.S. courts led to the clarification of three categories of *not-patentable subject matter.*

They are:

1. *The laws of nature,*
2. *Physical/Natural phenomena,* and
3. *Abstract ideas.*

These will be elaborated later in the chapter in greater detail. In addition, the Atomic Energy Act of 1954 excludes the patenting of inventions useful solely in the utilization of special nuclear material or atomic energy in an atomic weapon. See 42 U.S.C. 2181(a).[4]

According to the WIPO[5]:

> In order to be *eligible* for *patent* protection, an invention must fall within the scope of patentable subject matter. Patentable subject matter is established by statute and is usually defined in terms of the exceptions to patentability, *the general rule being that patent protection shall be available for inventions in all fields of technology.*

The TRIPS Agreement suggests to its members that *the following subject matter is non-patentable* (see the TRIPS Agreement, Annex 1C, Part II, Section 5, Article 27)[6]:

- *Discoveries* of materials or substances *already existing in nature;*
- *Scientific theories* or *mathematical methods;*
- *Plants and animals* other than microorganisms, and essentially biological processes for the production of plants and animals, *other than non-biological/microbiological processes;*
- *Schemes, rules, or methods,* such as those for doing business or performing *purely mental acts* or *playing games;*
- *Methods of treatment* for humans or animals, or *diagnostic methods* practiced on humans or animals (but not products for use in such methods); and finally
- The *commercial exploitation* of certain kinds of inventions *which would disregard public order or morality.*

Term of Patent Protection

In the USA, a *patent grant* will be for a term *beginning on the date on which the patent issues* and ending 20 years from the date of *filing of the patent application* (see 35 U.S.C. § 154).[7a]

Similarly, in the European Union (EU), a *patent is granted* for a period of 20 years from the *date of filing the patent application.*[7b]

In India, the Indian Patent Act (see IPA § 53) addresses the term of *patent* protection[7c] and provides for a term of 20 years from the filing date for every *patent* that is "granted" and "not expired or ceased to have effect," *after* the commencement of the *Patents (Amendment) Act, 2002,* that came into effect on May 20, 2003.[7d] Prior to that date, the terms of *patent* protection were different for process and product *patents* in India. Thus, the term of a *process patent* was 5 years from the date of sealing/7 years from the date of filing, whichever was earlier. On the contrary, the term of a *product patent* (for non-pharmaceutical/non-agrochemical products only) was 14 years from the date of the *patent*. It is noteworthy that India *did not allow the grant of a product patent for pharmaceutical products until* 2005. However, as India became a signatory to the TRIPS Agreement in 1995, it was given the transitional facility (mailbox) until January 1, 2005, for receiving and holding patent applications. Since then, the term of *patent* protection in India has become 20 years for both the *product* and *process patents* from the date of filing the application.

In China, the maximum term of protection for a *utility model patent* (UMs) is 10 years from the filing date, while an *invention patent* receives protection for 20 years.[7e]

Types of Patents in Different IP Regimes—A Brief Sketch

The USPTO grants *three different types of patents*[8]:

1. A *utility patent/invention patent* to anyone who invents or discovers any new and useful process, machine, article of manufacture, or composition of matter or any new and useful improvement thereof;
2. A *design patent* to anyone who invents a new, original, and ornamental design for an article of manufacture; and
3. A *plant patent* to anyone who invents or discovers and asexually reproduces any distinct and new variety of plant.[9]

In recent years, about 90% of the *patents* granted in the USA have been *utility patents. Design patents* issued from applications filed on or after May 13, 2015 shall be granted the term of 15 years from the date of grant, whereas those filed before May 13, 2015, shall be granted the term of 14 years. It is noteworthy that *utility patents* are subject to the payment of periodic maintenance fees to keep the *patent* in force, while *design patents* and *plant patents* are not subject to the payment of periodic maintenance fees.

In the EU, most countries grant only *invention patents.* According to the WIPO[10]:

> *Patents* are granted for *inventions in the field of technology,* from an everyday kitchen utensil to a nanotechnology chip. An *invention* can be a product—such as a chemical compound or a process—such as a process for producing a specific chemical compound. For example, a laptop computer can involve a hundred inventions, working together.

In contrast, China grants *three types of patents.* They are[11]:

1. An *invention patent* that is granted for an *invention,* namely, a new and inventive technical solution for a product, new method of producing or doing something, or an improvement to an existing

product. Once granted, *invention patents* are legally protected for 20 years from the application date.

2. A *utility model patent*—simply referred to as, UM—that is granted for *new and practical technical solutions* relating to the *shape and/or structure of a product*. These *patents* protect new, functional aspects of a product that do not meet the higher inventiveness level required for an invention *patent*. Once granted, UMs are legally protected for 10 years from the application date.
3. A *design patent* that is granted for *new designs relating to the shape, pattern, or their combinations*, or *the combination of color, shape, and/or pattern that are aesthetically pleasing and industrially applicable*. A *design patent* protects the "look" of the product that makes it recognizable. *Design patents* are protected for 10 years from the application date.

Besides China, the other countries that grant UMs include—Austria, Belgium, Denmark, Finland, Germany, Italy, the Netherlands, among others. However, the USA and United Kingdom do not have a UM system. The differences between *invention patents* and UMs in China are outlined in Table 3.1 below.[12]

Like the EU, India grants only *invention patents*. According to the Government of India: A *patent* is granted for an *invention* which is a *new* product or process involving an *inventive step* and *capable of industrial application*.[13] The Office

TABLE 3.1

Differences between *Invention Patents* and UMs

S. No.	Criteria	Description of the Differences
1	Subject matter	Unlike *invention patents*, UMs only protect products with new shape/structural physical features. Thus, methods of production or chemical compounds are not eligible for Ums, as they lack shape/structural physical features.
2	Inventiveness	Though *invention patents* are expected to possess a higher degree of "inventiveness" compared to UMs, in reality there is little difference between the two.
3	Examination	*Invention patents* require "substantive examination" that is very detailed (may take 3–5 years), while UMs need only "formality/preliminary examination" (may take up to a year).
4	Term of protection	The term of protection for an *invention patent* that is granted is 20 years from the filing date, while the maximum term of protection for a UM is 10 years from the filing date.
5	Cost	The application fee for an *invention patent* is RMB 900 while the same for a UM is RMB 500. The examination fee for an *invention patent* is RMB 2500, whereas there is no examination fee for a UM.

Source: script%20in%20Preparation/C-Chapters/Chapter%203/D-Reference%20Materials/China%20IPR%20Guide%20to%20Patent%20Protection%20in%20China%20(2013).pdf.

of the Controller General of Patents, Designs, and Trademarks, Government of India, defines the following[14]:

- *Invention* means a new product or process involving an *inventive step* and *capable of industrial application;*
- *Inventive step* means a feature of an invention that involves a technical advance as compared to the existing knowledge or having economic significance or both and that makes the invention not obvious to a person skilled in the art; and
- *Capable of industrial application,* in relation to an invention, means that the invention is capable of being made or used in an industry

Substantive Standards for Patentability—A Brief Sketch

According to the USPTO,[15a] an *invention* is *patentable* in the USA, *if it is:*

1. *New* (35 U.S.C. § 102);[15b]
2. *Useful* (35 U.S.C. § 101);[3e,15a] and
3. *Non-obvious* (America Invents Act (AIA) 35 U.S.C. § 103)[15c]

as defined by the U.S. *patent* laws listed above. This will be explained in greater detail later in this chapter.

Similarly, in the EU also, an *invention is patentable along practically the same lines.* According to the WIPO,[16] these key criteria are:

1. The invention must show an element of *novelty;* namely, a *new* characteristic *not known in the existing body of knowledge* in that specific technical field, known as, *prior art;*
2. The invention must involve an *inventive step* or it must be *non-obvious,* namely, it *cannot be obvious to a person having ordinary skill* in the relevant technical field;
3. The invention must be *capable of industrial application,* namely, *it must be capable of being used for an industrial or business purpose, beyond being a merely abstract phenomenon,* or *useful;* and
4. Its *subject matter* must be *patentable* under the law.

   Certain categories of subject matter—such as, scientific theories, aesthetic creations, mathematical methods, plant or animal varieties, discoveries of natural substances, commercial methods, methods for medical treatment (as opposed to medical products), or computer programs—are generally not patentable.

In India, *an invention must satisfy very similar conditions to the* USA *for grant of a patent.* According to the Indian Government[17]:

> A patent is granted for an invention which is a new product/process that meets the conditions of novelty, non-obviousness, and industrial use.

In China *an invention must also satisfy similar conditions for grant of a patent.* According to Article 22 of the Chinese Patent Act[18]:

> Any invention or *utility model* for which a *patent* right may be granted must have *novelty, inventiveness,* and *practical applicability.*[18]

Part-B: Patent Particulars

Patentable Subject Matter—Detailed Discussion

Analyzing the U.S. *patent* law is helpful here. As pointed out above, 35 U.S.C. § 101[3e] states:

> Whoever invents or discovers any new and useful process, machine, manufacture, or composition of matter, or any new and useful improvement thereof, may obtain a *patent* therefor, subject to the conditions and requirements of this title.

Thus, according to 35 U.S.C. § 101, the *patentable subject matter* is:

1. *New* and *useful* processes;
2. *New* and *useful* products (machine, manufacture, and composition of matter); and
3. *New* and *useful* improvements of processes/products.

Appropriate cases must be closely looked at to understand these categories.

Patentable Processes

According to 35 U.S.C. § 100 (b)[3c]:

> A "patentable process" is either a "new" *process, art,* or *method,* or a "new use" of a *known* process, machine, manufacture, composition of matter, or material.

The following clarification may be useful in this regard. A "process" is *the sequence of steps to achieve a goal.* A "method" is the *procedure to achieve a particular step* (or task). "Art" is the *technique to create something* or *achieve a useful result.* Further, a "process" can be a "method" *of making, using, or doing something.*

According to the case, *Gottschalk v. Benson*, 409 U.S. 63, 70, 175 USPQ 673, 676 (1972)[19a]:

> A process is a mode of treatment of certain materials to produce a given result. It is an *act*, or a *series of acts*, performed upon the subject-matter to be transformed and reduced to a different state or thing.[19b]

The following examples illustrate the point:

1. A novel chemical synthesis (the method of *making*, *new* process) of the anti-cancer drug "Taxol," involving a sequence of new reaction methods;
2. A new stage-wise application of laser technology (the method of *using*, *new* process) for effective cataract surgery;
3. A new step-wise process for accurate flight control of micro-drones (the method of *doing something*, *new* process) to detect injured victims in a collapsed building;
4. A new, efficient, and reproducible technique for Chinese woodblock printing (the art of woodblock printing, *new* art);
5. Thermal infrared imaging (*known* process) for precise mapping and detection of landmines over a large battle arena (*new* use);
6. Ordinary irrigation pump (*known* machine) for driving a farm tractor (*new* use);
7. Cellular phone (*known* article of manufacture) as an effective navigational tool in the dark (*new* use);
8. Sildenafil Citrate (*known* composition of matter, originally developed for treating arterial hypertension) for treating impotence (*new* use); and
9. Carbon nanotubes (*known* materials) for shielding and absorbing electromagnetic radiation (*new* use).

Patentable Products

According to 35 U.S.C. § 101, three categories—namely, *machines*, *articles of manufacture*, and *compositions of matter*—qualify as, *patentable products*. In the USA, the interpretation of this statute is very broad, and to gain understanding of this is critical:

Machine

According to the case, *Corning v. Burden*, 56 U.S. 252, 266, 267 (1853)[19c]:

> The term *machine* includes *every mechanical device or combination of mechanical powers* and *devices to perform some function and produce a certain effect*

> *or result.* But where the result or effect is produced by chemical action, by the operation or application of some element or power of nature, or of one substance to another, such *modes, methods,* or *operations* are called *processes.*[19c]

Speaking in the most basic terms, a "machine" *performs a useful operation.* Innumerable inventions fit the above description of a *machine.* Examples include:

1. Bicycle or "pedal velocipede" **U.S. 59915** (1866);
2. Portable harmonium **U.S. 1778885 A** (1930); and
3. Mouth argon **U.S. 2478963 A** (1949)

According to *Nestle-Le Mur v. Eugene,* 55 F. 2d 854, 857 (6th Circuit, 1932)[20a]:

> A *machine* is a device or combination of devices by means of which *energy can be utilized or a useful operation can be performed.* It is adapted to *rendering a mechanical service* or to *the fabrication of material* so as to change its form or produce a desired product.

Indeed, a large number of inventions fit this description of a *machine.* Such examples of *patented machines* include:

1. Electric fan **U.S. 1975934 A** (1934);
2. Electric iron **U.S. 259,054** (1882);
3. Morse's telegraph **U.S. 1,647** (1840);
4. Graham Bell's telephone **U.S. 174,465** (1876); and
5. Thomas Alva Edison's electric lamp **U.S. 223,898 A** (1880).

Indeed, 35 U.S.C. § 101 is interpreted broadly according to the Jefferson's philosophy that "ingenuity should receive a liberal encouragement. Patentable subject matter should include anything under the sun that is made by man.[20b]"

Thus, in the landmark case, *Diamond v. Chakrabarty,* 447 U.S. 303, 206 USPQ 193 (1980),[21] the U.S. Supreme Court examined whether a *genetically engineered micro-organism qualifies as a patentable invention* under 35 U.S.C. § 101. In a 5-4 decision, the U.S. Supreme Court ruled that while *the discovery of a natural law, natural phenomenon, abstract idea, or new mineral is not patentable* (as the subject matter is natural, abstract and/ "not made by a human"), *a genetically engineered micro-organism* (made by human), *which is not naturally occurring* (new and non-obvious), *and oil-eating* (useful), is *patentable* as "manufacture" and "composition of matter."

In this case, the U.S. Supreme Court noted that:

> [the U.S.] Congress thus recognized that the *relevant distinction was not between living and inanimate things, but between products of nature, whether*

> *living or not, and human-made inventions.* Here, respondent's micro-organism *is the result of human ingenuity and research*[22]

Accordingly, in 1987, the U.S. Commissioner of Patents and Trademarks issued a notice[23] which read:

> USPTO would consider *non-naturally occurring, non-human multi-cellular living organisms,* including animals, to be *patentable subject matter* within the scope of 35 U.S.C. § 101.

Manufacture

In the case, *Diamond v. Chakrabarty*, 447 U.S. 303, 206 USPQ 193 (1980),[21] the U.S. Supreme Court read "manufacture" as:

> An *article produced from raw or prepared materials* by giving to these materials new forms, qualities, properties, or combinations, whether by hand labor or by machinery.[24]

Indeed, according to this broad interpretation, numerous articles qualify as patentable "articles of manufacture." Illustrative examples are:

1. Breathable apparel/footwear/gloves manufactured from Gore-Tex® fibers;[25]
2. Diabetes-controlling bread handmade from a unique combination of five ancient grains;
3. Lightweight fuselage and wings of the Boeing 787 Dreamliner manufactured from weight-saving advanced composite materials; and
4. Anti-bacterial meat-cutting board manufactured from silver-impregnated wood.

It is important to point out that the newly invented elements *must be functional* for the "article of manufacture" to be *within the patentable subject matter.*

Composition of Matter

The category of "composition of matter" covers a wide range of inventions, such as the following, to mention a few:

- New *drug molecules* (i.e., new chemical compounds);
- New *synthetic materials* (i.e., nanomaterials, superconductors, plastics, etc.);
- New *formulations* of known drugs (i.e., significant improvements);
- New *drinks/beverages* (such as energy drinks, vitamin concoctions, etc.);
- New *food recipes*, including new flavoring substances;

- New *skin care formulations* (for enhanced skin benefits);
- New *air-cleaning formulations* (for reduction of air pollution);
- New *water-cleaning formulations* (for purer drinking water);
- New *home-cleaning formulations* (for improved hygiene);
- New *industrial-cleaning formulations* (for cleaner environment); and
- New *laundry detergent compositions* (for cleaner fabric wash).

In *Diamond v. Chakrabarty*, 447 U.S. 303, 308, 206 USPQ 193, 197 (1980), the "composition of matter" is defined as:

> All *compositions of two or more substances* and *all composite articles*, whether they be the results of chemical union, or of mechanical mixture, or whether they be gases, fluids, powders, or solids.

The following examples are illustrative:

1. A Voriconazole (*new drug molecule*) composition for treatment of a variety of anti-fungal infections (solid, chemical mixture) **U.S. 20140275122 A1** (2014);[26a]
2. A flame-resistant *composition of matter & formed articles* made from high molecular weight polyesters (solid, chemical mixture) **U.S. 3929720 A** (1975);[26b]
3. Digestion-enhancing *curry powder* made from a combination of new herbs and spices (powder, mechanical mixture); and
4. Wrinkle-reducing, moisturizing *skin lotion* made from oils, fats, dye, perfume, water, and new benefiting agents (liquid, mechanical mixture).

It is important to underscore that this category is interpreted very broadly. Most notably, a *genetically engineered, living micro-organism* is allowed as a patentable "composition of matter" in *Diamond v. Chakrabarty*, 447 U.S. 303, 308, 206 USPQ 193, 197 (1980).

Patentable Improvements

According to 35 U.S.C. § 101, *new and useful improvements* of processes and products (*machine, article of manufacture,* and *composition of matter*)—are also considered as *patentable subject matter.* The following examples are illustrative:

1. Improvement of the *process* of curing synthetic rubber using continual temperature measurements and programmed digital computer in several steps. See, *Diamond v. Diehr* 450 U.S. 175 (1981);[27a]
2. Improvement of a *sewing machine* **U.S. 8294 A** (1851);[27b]

3. Improvement of a *type-writing machine* **U.S. 79265 A** (1868)[27c];
4. Improvement in *phonograph* or *speaking machine* **U.S. 200521 A** (1878)[27d];
5. Improved *composition of matter*, called *Metaline* **U.S. 101866 A** (1870)[27e];
6. Improvement in the *composition of matter* for firebricks **U.S. 393 A** (1837).[27f]

There is one more point we need to mention here, and that is the issue of *blocking patents*. Suppose, inventor-A invents a new and useful Object-A (i.e., a ceramic mug), and inventor-B invents new and useful improvement to that Object-A (i.e., a ceramic mug with a handle that helps in holding the mug), then both the *original invention* (a ceramic mug) and the new and useful *improvement* (ceramic mug with handle) may receive *patents* (provided, of course, they also satisfy the other conditions of 35 U.S.C. § 101). Here, we need to remind ourselves that the *exclusive right* of a *patent* is a *negative exclusionary right* and not a *positive affirmative right*. In practical terms, what that means is that though both inventor-A and inventor-B have *patent* rights, neither can practice the *improvement of* Object-A (ceramic mug with handle), *as they block each other* from practicing the invention. In such situations, the inventors may consider licensing/cross-licensing/buying of each other's *patents*.

Not-Patentable Subject Matter—Detailed Discussion

As pointed out in the case *Diamond v. Chakrabarty*, 447 U.S. 303, 308, 206 USPQ 193, 197 (1980), *the laws of nature, physical phenomena*, and *abstract ideas do not qualify as patentable subject matter.*

Further, the purpose and goals for the protection of *intellectual property* are *simple*, *straightforward*, and *positive*. The WIPO nicely summarizes this as follows:

> In a wider sense, *the public disclosure of the technical knowledge in the patent*, and *the exclusive right granted by the patent*, provide incentives for competitor to search for alternative solutions and to "invent around" the first invention. These incentives and the dissemination of knowledge about the new inventions *encourage further innovation, which assures that the quality of human life and the well-being of society is continuously enhanced.*[28]

Thus, one must recognize that the *outcomes of innovations* can be either "useful" or "harmful" to society. Therefore, nations interested in promoting *the quality of human life and the well-being of society* are careful about granting *patent* protection to only those inventions that are useful to mankind and not to immoral inventions or inventions harmful to mankind. The following illustrative examples can be helpful in this regard.

Laws of Nature

At the heart of this debate is the idea of *discovery vs. invention*. Both involve application of human intellect and human effort. However, a *discovery* involves *uncovering a pre-existing truth,* while an *invention* involves *creation of something new or improved through human effort*. Thus, one can see why a discovery is "not patentable," however impactful it might be, while an invention "is" (which involves human ingenuity for making something new or improved). Indeed, this is the reason why even *Newton's Findings on Motion and Gravity* are "not patentable"—as they are NOT *inventions* and ONLY ingenious *discoveries* of natural (pre-existing) truths.

Physical/Natural Phenomena

Physical/natural phenomena *are not-patentable, however significant or useful they might be*. For instance, a scientific finding that *plants conduct photosynthesis* may greatly advance science, but is *not patentable,* as it is a *discovery* and not an *invention*. Similarly, a (hypothetical) scientific finding that certain *naturally occurring bacteria consume large amounts of carbon dioxide* will be *not patentable,* as it is also a *discovery*. Further, the *products of nature are not patentable as they too constitute discoveries of natural phenomena*. For example, a *new chemical element found in the volcanic ash* of Hawaii or a *new species of butterflies* found in the Amazon jungle—are *not patentable* as they are *discoveries* of natural phenomena. Similarly, if a scientist—who studies the exhaust emissions of cars—*finds* that certain brands of cars emit more polluting smoke compared to the rest, then such a finding will *not be patentable,* as it is *a discovery* of physical phenomenon and not an *invention*. See, *Cf. Nippon Electric Glass Co. v. Sheldon,* 539 F. Supp. 542 (S.D.N.Y. 1982).

Abstract Ideas

"Pure mathematics" and "abstract ideas" are *not patentable*. The relevant inventions here are: *computer software* and *business methods*. The interpretation of *patent* law is *very tricky with respect to computer software*. Indeed, certain landmark cases (shown below) highlight the intricate issues and provide critical guidance in this regard. In early cases (see the first two cases below), *computer software* was generally considered as a mathematical algorithm/abstract principle *not patentable*. However, in later cases (see the next four cases below), the court rulings suggest that *software* and *business methods are patentable under certain conditions*.

In *Gottschalk v. Benson,* 409 U.S. 63 (1972),[29a] Benson (Plaintiff) submitted *patent* claims for *a method of converting binary coded decimal numbers into pure binary numbers*. This was not specific to any methodology, equipment, or machinery, or even end use. Nevertheless, their goal was to get a broad *patent* coverage of their *mathematical method* for all uses of including even use of it by a human with a pencil and paper. The *patent* examiner rejected their invention as not patentable based on the grounds that it is an *abstract principle* or a *mathematical algorithm* (a method that solves a mathematical problem). However, an appellate court (the Court of Customs and Patent Appeals, CCPA) *overruled* the *patent* examiner and *ordered a patent to issue.* Then, Gottschalk (defendant), the Acting Commissioner of Patents, successfully petitioned the U.S. Supreme Court to review the appellate court's decision. The two key issues were: (a) is a *computer program* patentable? and (b) is a *mathematical algorithm* patentable? The Supreme Court answered "no" and "no" to both the questions in this case.

In a 6-0 ruling, the U.S. Supreme Court *reversed* the CCPA's decision, and stated:

> Respondents' *method for converting numerical information from binary-coded decimal numbers into pure binary numbers,* for use in programming conventional general purpose digital computers, *is merely a series of mathematical calculations* or *mental steps* and *does not constitute a patentable process* within the meaning of the Patent Act, 35 U.S.C. § 100(b). Pp. 409 U.S. 64–73.

In *Parker v. Flook,* 437 U.S. 584 (1978),[29b] Dale Flook (plaintiff) submitted *patent* claims for a *method of adjusting alarm limits in response to changes that occur during catalytic conversion.* Catalytic conversion is a step in the oil-refining process for the removal of pollutants, which needs to operate within certain conditions, such as—temperature, pressure, and flow rates (the "alarm limits")—that need continual monitoring to be effective. The *patent* examiner rejected the invention *as not patentable* on the grounds that the only novelty in the *patent* application is a *mathematical formula.* Next, the Board of Appeals for the U.S. Patent and Trademark Office sustained the rejection. On appeal, the Court of Customs and Patent Appeals *reversed* the Board of Appeals decision and held that the *invention was eligible for patent protection.* Then, Parker (defendant), the Acting Commissioner of Patents, petitioned the U.S. Supreme Court to review the appellate court's decision, urging that CCPA's decision will have a devastating effect on the rapidly growing computer "software" industry.

In a 6-3 decision, the U.S. Supreme Court *reversed* the appellate court's decision and stated:

> The respondent's *method for updating alarm limits during catalytic conversion processes,* in which the only novel feature is a *mathematical formula,* held *not patentable* under § 101 of the Patent Act. The identification of a limited category of useful, though conventional, post-solution applications of such a formula *does not make the method eligible for patent protection,* since, assuming the formula to be within prior art, as it must be, O'Reilly v. Morse, 15 How. 62,[29g] respondent's application *contains no patentable invention.* The chemical processes involved in catalytic conversion are well known, as are the monitoring of process variables, the use

> of alarm limits to trigger alarms, the notion that alarm limit values must be recomputed and readjusted, and the use of computers for "automatic process monitoring."

Thus, it appears that the above two cases suggest that *computer software* and *mathematical formulae are not patentable.* However, the following cases suggest that *computer software may be patentable* under certain conditions.

In *Diamond v. Diehr*, 450 U.S. 175 (1981),[29c] Diehr (plaintiff) invented *a process for curing synthetic rubber that depended on a mathematical formula and a programmed computer* in its multiple steps.[29h] The *invention* involved the process of curing rubber by constantly measuring some specific temperatures, feeding the information to a computer, which then *would calculate (based on a mathematical algorithm or computer software) the precise time to terminate the curing process.* Considering computer software "unpatentable" based on the earlier Supreme Court decisions (see the above two cases), the *patent* examiner *rejected* Diehr's invention under 35 U.S.C. § 101, citing that the invention simply combined *an unpatentable computer software with a conventional rubber curing process.* This case was then appealed to the appellate court (CCPA). The appellate court (CCPA) *reversed* the *patent* examiner's decision and *ordered the patent to be issued.* Diamond (defendant), the Acting Commissioner of Patents, petitioned the U.S. Supreme Court to review the appellate court's decision.

In a 5-4 decision, the U.S. Supreme Court *held* the appellate court's decision and stated that:

> When a claim containing a *mathematical formula implements or applies the formula in a structure or process which, when considered as a whole, is performing a function which the patent laws were designed to protect* (e.g., transforming or reducing an article to a different state or thing), *then the claim satisfies § 101's requirements.* Pp. 450 U.S. 191–193.[29i]

In essence, the Supreme Court allowed the patenting of an *improved process* (for curing rubber) *made possible, of course, by a computer software* (the only novel element in the invention), which is within the subject matter of 35 U.S.C. § 101.

The Supreme Court *issued no further opinions on the patentability of computer software, leaving the interpretation of the law to the Federal Circuit.* Accordingly, three famous Federal Circuit opinions—namely, *In re Allapat*, 33 F.3d 1526 (Fed. Cir. 1994),[29d] *State Street Bank & Trust v. Signature Financial Group*,[29e] 149 F.3d 1368 (Fed. Cir. 1998), and *AT&T Corp. Excel Communications*, 172 F.3d 1352 (Fed. Cir. 1999)[29f]—play a useful role in *showing how patent eligibility is determined* in certain cases of *software patents* and *patent applications*. These opinions opened a floodgate of software and business-method *patent* applications.

In *In re Alappat*, 33 F.3d 1526 (Fed. Cir. 1994),[29d] the *patent* application describes an invention that involves a *machine* ("rasterizer") *which performs a sequence of steps, according to the instructions generated by a software* (mathematical algorithm), loaded on a digital computer. To start with, the *patent* examiner rejected Alappat's claims as *non-statutory subject matter* under 35 U.S.C.

§ 101. Then, Alappat appealed the case to the Board of Patent Appeals and Interferences (BOPAI). The 3-member panel of BOPAI *dissented* with the *patent* examiner's decision stating that the invention falls within the statutory subject matter of 35 U.S.C. § 101, as the claim recites a *means-for performing function, as it was directed to a machine*. Then the *patent* examiner requested for reconsideration of the 3-member BOPAI panel decision by an expanded panel, stating that the decision conflicted with PTO's policy. Interestingly, the 8-member expanded panel of BOPAI struck down the 3-member BOPAI panel's decision and *affirmed* the original *patent* examiner's rejection of the claim that a computer program is unpatentable under 35 U.S.C. § 101 because it recited a mathematical algorithm. At this stage, Alappat appealed the BOPAI expanded panel's rejection to the Court of Appeals for the Federal Circuit (CAFC). On the merits of Alappat's *patent* application, CAFC finally *reversed* the BOPAI expanded panel's decision and held that *the use of a mathematical algorithm in software does not bar a machine from patentability on Section 101 subject matter grounds*. In doing so, CAFC clarified its position by noting that *the invention in question was not an unpatentable abstract idea*, but a specific *machine* ("rasterizer") that was made to produce a *useful, concrete, and tangible result under instructions given by a software* (mathematical algorithm).

In *State Street Bank & Trust v. Signature Financial Group*, 149 F.3d 1368 (Fed. Cir. 1998),[29e] Signature Financial Group developed *a data processing system* to implement an investment structure *for efficient administration of large-scale portfolios of mutual funds*. This *investment structure*—identified by the proprietary name *Hub and Spoke®*—provides a system whereby mutual funds (Spokes) pool their assets in an investment portfolio (Hub) organized as a partnership. Thus, the *Hub and Spoke®* system provides the mutual fund administrator economies of scale in administering investments, as well as tax advantages of a partnership. The Signature Financial Group received the *patent* U.S. 5,193,056 (1993) for this *computer-implemented business method*, which will be simply referred to as the '056 *patent* from now on. This '056 *patent* was drafted along the lines of *Diamond v. Diehr*, 450 U.S. 175 (1981). It is important to note that at first, State Street and Signature were custodians and accounting agents for multitiered partnership fund financial services. Later, State Street Bank tried to license the '056 *patent* from the Signature Financial Group. But, when their negotiations broke down, State Street Bank brought an action seeking a declaratory judgment in Massachusetts district court that the '056 *patent* was *invalid* under 35 U.S.C. § 101, as it claimed an *unpatentable mathematical algorithm*. The district court considered the central issue in this case to be, whether *computer software* (a mathematical algorithm)—*which gave instructions to a computer that performed mathematical and accounting functions* (business method)—was "patentable subject matter" under 35 U.S.C. § 101. In the first part of the opinion, the district court ruled that the '056 *patent* constituted an *unpatentable mathematical algorithm* and declared the *patent invalid*. To arrive at this conclusion, the district court employed the

Freeman-Walter-Abele test (simply referred to as, the "Freeman test")[30a–c] as a guidepost to determine the patentability of an invention involving computer software (mathematical algorithm). According to the court, the invention failed to meet the *Freeman test standard,* as it did not involve *the transformation or conversion of subject matter representative of or constituting physical activity or objects* and only involves a change of one set of numbers into another, *which makes it unpatentable.* In the second part of the opinion, the district court concluded that the '056 *patent qualified as an unpatentable business method,* as the invention fell within the scope of "business plans" and "systems" (abstract ideas), which have been considered as *unpatentable for a long time.*[30d] At this stage, Signature Financial Group *appealed* the case to the Federal Circuit.

The Federal Circuit decision clarified important aspects of this case:

1. First, the court concluded that the technical description of the '056 *patent—with its means plus function claims structure*—fully supported the conclusion that *the patent claims and their equivalents described a machine.*

   For this, the court clarified its position by saying that:

   > patentability depended on whether the invention fell within one of the four categories of 35 U.S.C. § 101, namely—*machine,* article, *composition of matter,* or process…

2. Second, the court engaged in an extensive statutory analysis of 35 U.S.C. § 101, and articulated its insights, and concluded that:

   > the plain and unambiguous meaning of [35 U.S.C.] § 101 is that *any invention falling within one of the four stated categories* of statutory subject matter may be patented…" The court emphasized the intent of the law by further pointing out that: "the Supreme Court had previously decided that Congress intended [35 U.S.C.] § 101 *to extend to anything under the sun that is made by man.*[21]

3. The Federal Circuit noted that the Supreme Court in *Diamond v. Chakrabarty,* 447 U.S. 303, 308, 206 USPQ 193, 197 (1980), had opined that *mathematical algorithm is not patentable subject matter only when it is just an abstract idea.* The Federal Circuit concluded that *a mathematical algorithm is patentable, when the algorithm controls a practical application, to produce a useful, concrete, and tangible result.*[29d]
4. Advising that the inquiries into the patentability of inventions should concentrate on the essential characteristics of the subject matter, namely, its *practical utility* (namely, whether the invention produces a *useful, concrete, and tangible result*), the Federal Circuit took the opportunity and discredited the *Freeman-Walter-Abele test.*

5. The court finally turned its attention to the judicially created "business method" exceptions to subject matter.[30d] The Federal Circuit noted that:

> the *doctrine of business method exception* merely represented the application of some general, but no longer applicable legal principle... which was eliminated by [35 U.S.C.] § 103, including non-obvious subject matter.[31]

In essence, based on the reasoning outlined above, the Federal Circuit court *reversed* the Massachusetts district court's decision by ruling that *computer software* (a mathematical algorithm)—which gave instructions to a computer that performed mathematical and accounting functions (business method)—*that produced a useful, concrete, and tangible result was patentable subject matter under* 35 U.S.C. § 101.

In *AT&T Corp. v. Excel Communications*, 172 F.3d 1352 (Fed. Cir. 1999),[29f] AT&T sued Excel Communications, Excel Communications Marketing, and Excel Telecommunications (collectively "Excel") for infringement of ten of the method claims of its *patent*, U.S. 5,333,184 (from now will be referred to as "the 184 *patent*").[32a,b] The "184 *patent*" describes an improved technology (computer software/mathematical algorithm) to calculate the price of direct-dialed long-distance telephone calls, which telephone service providers can use to generate customers' bills (business method). Thus, the "184 *patent*" allowed AT&T to identify its own subscribers and offer a discount on calls to other AT&T subscribers. Excel moved for summary judgment on the grounds that "the 184 *patent*" did not meet the statutory requirements of patentable subject matter under 35 U.S.C. § 101.[32b] The district court granted Excel's summary judgment motion. First, the court determined that the claims at issue implicitly recite *a mathematical algorithm*. Second, the court tried to establish "whether the process claimed in the '184 *patent' performed a function*." On the grounds that *patentable processes must transform the substance of input data, and not merely the form*, the court ruled that the claims of the "184 *patent*" were *invalid* under 35 U.S.C. § 101. At this point, AT&T appealed the case to the Federal Circuit. Interestingly, the Federal Circuit *upheld the method claims* of the "184 *patent*" against the 35 U.S.C. § 101 challenge. Judge Plager, writing for the unanimous panel, justified the decision on the grounds that "*patent* law must adapt to new and innovative concepts, while remaining true to basic principles." Further, the judge refused to distinguish *Excel* from *Alappat* and *State Street Bank* (the cases we already discussed above), both of which involved *machine claims*, holding instead that the court considers "the scope of section 101 to be the same regardless of the form—*machine* or *process*—in which a particular claim is drafted." In continuation, the Federal Circuit ruling stated: Because section 101 states that processes can be patented, *any proscription against patenting algorithms must be narrowly limited to mathematical algorithms in the abstract*. In earlier rulings, the Federal Circuit rejected process claims involving a mathematical algorithm unless they encompassed a physical transformation.[32c] However, in this case, it upheld process claims of the "184 *patent*" stating that a physical transformation was *not an invariable requirement, but merely one example* of how an algorithm can be applied to achieve a *useful, concrete, and tangible result*. In conclusion, the Federal Circuit ruled that:

> AT&T's claimed process (business method) was a *useful application* of *Boolean algebra* (mathematical algorithm) and that *the process therefore comfortably falls within the scope of* 35 U.S.C. § 101."

Immoral Inventions

It is very important to point out that this is *not a category of expressly prescribed exclusions*. However, as pointed out in the very beginning of this section, one must recognize that innovation outcomes can be either *helpful* or *harmful* to society. Therefore, promotion of *the quality of human life and the well-being of society* requires careful granting of patent protection to only those *inventions that are useful to mankind* and not to *immoral inventions* or *inventions harmful to mankind*. The following illustrative examples can be helpful in this regard.

In the USA, the Congress and USPTO stipulate inventions that *embrace a human being* or a *method of cloning a human*—as unpatentable. Similarly, Great Britain excludes the following as unpatentable subject matter: *processes for cloning humans; modifying the germ line genetic identity of humans; uses of human embryos for industrial or commercial purposes*; and *processes for modifying the genetic identity of animals* which are likely to cause them suffering without any substantial medical benefit, and *animals resulting from such processes*. Along the same lines, in France, inventions that are *contrary to the dignity of the human person, public policy*, or *morality* are excluded from patentable subject matter. Similarly, China excludes invention-creation that is contrary to the laws or social morality or is *detrimental to public interest*. Finally, India does not allow inventions *relating to atomic energy* which fall within § 20(1) of the Atomic Energy Act 1962. Similarly, in the USA, the Atomic Energy Act of 1954 excludes the patenting of inventions in the utilization of special nuclear material or atomic energy—useful solely in an atomic weapon. Indeed, in any country, inventions to create the *weapons of mass destruction*—whose sole intent and purpose is mass extinction—are *unpatentable*, as national security and welfare of mankind take priority over *patent* law.[33]

Substantive Standards for Patentability—Detailed Discussion

As we have seen up to this point, an invention qualifies itself as "*patent* eligible" only if the *subject matter* of the invention falls *within the scope of the patentable subject matter*. After this procedural requirement, there are *three substantive standards* that an invention must satisfy for patentability. See discussion in Part-A, above. Accordingly, the *invention* must be:

1. *New* (35 U.S.C. § 102);[15b]
2. *Useful* (35 U.S.C. § 101);[3e,15a] and
3. *Non-obvious* (AIA 35 U.S.C. § 103).[15c]

Let us now discuss each of the above requirements in detail, one at a time.[34]

New—Novelty and Prior Art

This is the first *substantive requirement*. Under 35 U.S.C. § 102, an *invention* must satisfy the following *statutory conditions* to be considered "new":

Anticipation

According to 35 U.S.C. § 102 (a) & (e), the invention must NOT be already:

- known to the *public*;
- in *public use*/in sale *in the United States;*
- described in a printed publication *anywhere in the world*; or
- described in *a pending U.S. patent application.*

CLARIFICATION AND EXAMPLES

The central idea is that an *invention* in a *patent* application *cannot be considered* "new," if it is:

1. *already known to public;*
2. *already in public use or in sale in USA;*
3. *published in any form anywhere in the world that has public access to* (which constitutes *anticipation* by the "open prior art"); or
4. *described in a patent application pending with the USPTO* (which constitutes anticipation by the "secret prior art") *before the effective filing date of the claimed invention.*

The following clarifications will be helpful.

- ***Already known to public/in public use/in sale in USA?*** 35 U.S.C. § 102 (a). Suppose, an inventor in the USA believes that he/she invented a *novel method for making* an umbrella. However, he/she soon recognizes that many residents of the neighboring town also have been making the same exact umbrellas even earlier and using them as well. Clearly, the invention now

(*Continued*)

CLARIFICATION AND EXAMPLES (*Continued*)

constitutes *common knowledge* in the USA. Unfortunately, the inventor can no longer *patent* his invention because it is *already known to public* and *in public use* in the USA. Further, the invention will also be *unpatentable* if the umbrella were to be *already in sale.*

- ***In printed publication anywhere in the world?*** 35 U.S.C. § 102 (a). If inventor-A's *invention* (product/process) is *described by others in any kind of printed publication anywhere in the world* – such as, (a) the PhD thesis located in a university library in another country, (b) in a journal article, (c) in a book, or (d) even a newspaper article (regardless of how wide its circulation)—it is *unpatentable* (even if inventor-A is genuinely unaware of it!). However, it is very important to point out that *every element of inventor-A's invention* and all the information necessary to make or use the claimed invention *must be described in a single reference or prior art* for it to be disbarred from patenting. On the other hand, suppose an inventor-A invents a new process and submits a *patent* application to the USPTO. Let us also suppose that the exact same process has also been invented by another inventor-B, who *ONLY records the invention in a private research notebook* and *never discloses it to anyone nor publish it in any form anywhere.* Under such a scenario, inventor-B's invention privately recorded in his/her own research note book would not constitute "open prior art." Accordingly, 35 U.S.C. § 102 (a) *would not bar* inventor-A from patenting his/her process, as inventor-A invented it "independently" and he/she *is also the first one to file the patent application* with USPTO (The Leahy-Smith *America Invents Act* is in effect since March 16, 2013).[35]
- ***Described in a pending patent application?*** 35 U.S.C. § 102 (e). When an invention (product/process) *claimed in a newly submitted patent application is identical to the one described in a patent application that is already under review* by the USPTO ("secret prior art"), then it cannot be patented, as it is anticipated by the "secret prior art" (though the new *patent* applicant has no way of knowing about it!).

Derivation

According to 35 U.S.C. § 102 (f), the *inventor alone is entitled to patent his/her invention. No one else*—who might learn about the invention from the inventor or the others—*can patent an invention that they did not invent, even if the inventor has no intention of patenting his/her invention*. In fact, the derivation statute is so clear that even when the inventor assigns his/her rights to someone else, the *assignee must correctly identify the actual inventor* in the *patent* application.

Useful—The Utility Requirement

This is the second *substantive requirement*. For an invention to be *patentable,* it must be *useful*. See, 35 U.S.C. § 101. However, the utility requirement is quite *modest* and *can be met* if the invention could demonstrate an *identifiable benefit*.

CLARIFICATION AND EXAMPLES

To start with, an invention need not be *socially beneficial,* but it must possess *an identifiable benefit*. However, *this benefit cannot be merely described* by the inventor and *must be demonstrated through acceptable testing*. For instance, let us say an organic chemist isolated a chemical substance from the leaves of a tree by a "new" process, purified it, determined its structure by a combination of physical methods (such as NMR,[1] FTIR,[2] UVVIS,[3] and X-ray crystallography) and *found that its structure resembles that of other alkaloids which possess anti-malarial properties*. Let us assume that the inventor then applied for a U.S. *patent* on this invention, *claiming* "novelty" of the chemical process, and making the case for *its utility* (on the grounds that the isolated compound also will have anti-microbial properties because other similar structures possess anti-malarial properties), *without actually proving it through acceptable testing protocol*. Indeed, *this description alone will be insufficient to meet the utility standard of 35 U.S.C. § 101, without the actual proof,* however logically it may be argued. The benefit must be *specific, substantial,* and *practical*. See, the case of *Brenner v. Manson*, 383 U.S. 519 (1966).[36]

Further, the *identifiable benefit cannot be claimed merely on aesthetic grounds* and one must show that the invention actually works. Notably, the utility of an invention can be even worse compared to a benchmark. However, it is sufficient if it can be shown that it works. In fact, *an invention may meet the utility requirement even when it is deceptive*, as long as it shows an *identifiable benefit*. This is a tricky point. The case of *Juicy Whip v. Orange Bang*. 185 F. 3d 1364 (Fed. Cir. 1999) is illustrative.[37a,b] In this case, Juicy Whip, Inc., obtained the *patent* U.S. 5,575,405 for a *Post-Mix Beverage Dispenser with an Associated Simulated Display of Beverage*. The invention is

[1] Nuclear magnetic resonance spectroscopy
[2] Fourier transform infrared spectroscopy
[3] Ultraviolet visible spectroscopy

(*Continued*)

CLARIFICATION AND EXAMPLES (*Continued*)

designed such that the post-mix beverage dispenser is made to look like a pre-mix beverage dispenser, though it stores beverage syrup concentrate and water in separate compartments until the beverage is ready to be dispensed. Thus, when the consumer requests for beverage, the *machine* automatically mixes the syrup and water immediately and dispenses the beverage. The claims for the post-mix dispenser include a transparent bowl that is filled with an imitation fluid that gives the perception of the real beverage ("deceptive"). In actual fact, the beverage is mixed immediately prior to it being dispensed, as in conventional post-mix dispensers, which eliminates the need for frequent cleaning and refilling of the beverage compartment due to bacterial growth (a problem that plagues the pre-mix dispensers, a valuable improvement in actual reality).

The case facts are as follows: Juicy Whip sued Orange Bang, Inc., and Unique Beverage Dispensers, Inc., (collectively, the defendants, "Orange Bang") for infringement of its *patent* U.S. 5,575,405, in the United States District Court for the Central District of California. Then, Orange Bang moved for summary judgment of invalidity. The district court granted Orange Bang's motion and held the '405 *patent invalid* for lack of utility on the grounds that the *patented invention* (post-mix dispenser) *was designed to deceive* customers by imitating another product (pre-mix dispenser) to increase the sales of a particular good and *was thus unpatentable* under 35 U.S.C. § 101. Juicy Whip then appealed this decision to the U.S. Court of Appeals for the Federal Circuit, which *reversed* and *remanded* the decision of the U.S. District Court for the Central District of California, basing its arguments on the following grounds[40a]:

> It is not at all unusual for a product to be designed to appear to viewers to be something it is not. For example, cubic zirconium is designed to simulate a diamond, imitation gold leaf is designed to imitate real gold leaf, synthetic fabrics are designed to simulate expensive natural fabrics, and imitation leather is designed to look like real leather. In each case, the invention of the product or process that makes such imitation possible has "utility" within the meaning of the *patent* statute, and indeed there are numerous *patents* directed toward making one product imitate another. See, e.g., U.S. Pat. No. 5,762,968 (method for producing imitation grill marks on food without using heat); U.S. Pat. No. 5,899,038 (laminated flooring imitating wood); U.S. Pat. No. 5,571,545 (imitation burger). Much of the value of such products resides in the fact that they appear to be something they are not. Thus, in this case, the claimed post-mix dispenser meets the statutory requirement of utility by embodying the features of a post-mix dispenser, while imitating the visual appearance of a pre-mix dispenser.

Nonobvious—The Ultimate Condition of Patentability

This is the third *substantive requirement.* Non-obviousness requirement is considered as the "ultimate condition of patentability." According to the AIA 35 U.S.C. § 103:

> A *patent* for a claimed invention *may not be obtained,* notwithstanding that the claimed invention is not identically disclosed as set forth in section 35 U.S.C. § 102, if the differences between the claimed invention and the prior art are such that the claimed invention as a whole would have been **obvious before the effective filing date of the claimed invention** *to a person having ordinary skill in the art to* which the claimed invention pertains. Patentability shall not be negated by the manner in which the invention was made.

The USPTO explains the fundamental differences between the (pre-AIA) 35 U.S.C. § 103 and the AIA 35 U.S.C. § 103, in a detailed manner, as follows[38,39]:

> The AIA 35 U.S.C. 103 continues to set the non-obviousness requirement for patentability. There are, however, some important changes from pre-AIA 35 U.S.C. 103.
>
> The most significant difference between the AIA 35 U.S.C. 103 and pre-AIA 35 U.S.C. 103(a) is that the AIA 35 U.S.C. 103 determines obviousness **as of the effective filing date of the claimed invention**, rather than **as of the time that the claimed invention was made**. Thus, as a practical matter during examination, this distinction between the AIA 35 U.S.C. 103 and pre-AIA 35 U.S.C. 103 will result in a difference in practice only when the case under examination is subject to pre-AIA 35 U.S.C. 103 and there is evidence in the case concerning a date of invention prior to the effective filing date. Such evidence is ordinarily presented by way of an affidavit or declaration under 37 CFR 1.131.
>
> Next, the AIA 35 U.S.C. 103 differs from that of pre-AIA 35 U.S.C. 103 in that the AIA 35 U.S.C. 103 requires consideration of "the differences between the claimed invention and the prior art," while pre-AIA 35 U.S.C. 103 refers to "the differences between the subject matter sought to be patented and the prior art." This difference in terminology does not indicate the need for any difference in approach to the question of obviousness. As pointed out by the Federal Circuit, "[t]he term 'claims' has been used in *patent* legislation since the Patent Act of 1836 to define the invention that an applicant believes is patentable." *Hoechst-Roussel Pharms. Inc. v. Lehman*, 109 F.3d 756, 758, 42 USPQ2d 1220, 1222 (Fed. Cir. 1997) (citing Act of July 4, 1836, ch. 357, § 6, 5 Stat. 117). Furthermore, in *Graham v. John Deere*, the second of the Supreme Court's factual inquiries (the "Graham factors") is that the "differences between the prior art and the claims at issue are to be ascertained." 383 U.S. 1, 17, 148 USPQ 459, 467.
>
> Thus, in interpreting 35 U.S.C. 103 as enacted in the 1952 Patent Act—language that remained unchanged until enactment of the AIA—the court equated the subject matter sought to be patented with the claims.

CLARIFICATION AND EXAMPLES

The Supreme Court case of *Graham v. John Deere*, 383 U.S. 1 (1966), describes the steps of *non-obviousness analysis* as follows[40]:

1. Learn *the scope and content* of the prior art;
2. Ascertain *the differences between the claimed invention* and *the prior art;*
3. Determine *the level of ordinary skill* in the pertinent art; and
4. Establish the *obviousness/non-obviousness* of the claimed invention by: (i) analyzing the above three determinations, (ii) considering "secondary factors," such as the evidence of commercial success, long-felt, but unsolved needs, failure of others, and unexpected results, as well as (iii) applying the teaching, suggestion, and motivation (TSM) test.

Let us look at each of these separately.

1. ***Learn the scope and content of the prior art.*** It is important to point out that the AIA 35 U.S.C. § 103 does not identify the prior art that courts ought to consider for deciding whether an invention is obvious. Instead, this statute says that an invention is *unpatentable* if it is determined that: (a) the invention would be *obvious to a person of ordinary skill in the categories of prior art identified by* 35 U.S.C. § 102 and (b) *before the effective filing date of the claimed invention.* Therefore, for non-obviousness analysis, courts look to the prior art categories identified by 35 U.S.C. § 102. Another important criterion is that the prior art must be "reasonably pertinent" to the invention being considered under § 103.

   Non-obviousness analysis involves many steps. To do so, a *patent* examiner first looks at the nature of the problem being tackled by the inventor. Next, the examiner determines whether the references (prior art) fall within the inventor's field of endeavor. Then, he asks whether the references are "reasonably pertinent" to the invention being considered. See, *In re Paulsen*, 30 F.3d 1475, 1481 (Fed. Cir. 1994).[41] Then, the examiner follows the steps described below: ascertain the differences between the prior art and the claimed invention, determine whether it would be obvious to a person of ordinary skill in the art, and, finally, establish whether the invention is patentable and the claims allowable.

(*Continued*)

CLARIFICATION AND EXAMPLES (*Continued*)

2. ***Ascertain the differences between the prior art and the claimed invention***. The differences between the claimed invention and prior art could be of two types. In the first case, all elements of the claimed invention may not appear in a single reference, but are available in multiple references. In this scenario, the *patent* examiner must decide whether a person of ordinary skill in the art could connect them all together and make the same invention. In the second case, all elements of the claimed invention may not be available in the prior art. In this scenario, the examiner must decide whether a person of ordinary skill in the art could easily figure out the missing elements and make the same invention.
3. ***Determine the level of ordinary skill in the art***. The question is how does one decide the level of ordinary skill in the pertinent art? There are several relevant factors that one must consider in this regard, such as: (a) what are the problems encountered in the art? (b) Are there any prior art solutions to those problems? (c) What is the pace of innovations in this field? (d) How sophisticated is the technology? and (e) what is the literacy level of workers actively engaged in this field? The important question then is that "would the claimed invention be obvious to a person of ordinary skill in the art before its effective filing date?"
4. ***Establish the obviousness/non-obviousness of the claimed invention***. The case of *KSR International Co. v. Teleflex Inc. (KSR)*, 550 U.S. 398, 82 USPQ2d 1385 (2007) is illustrative in this regard.[42] In this case, the Supreme Court's decision reinforced earlier decisions and validated a more flexible approach to establishing the obviousness of an invention.

   The case facts are as follows: Teleflex, Inc. (plaintiff) brought a lawsuit against KSR International Co. (KSR) (defendant) in a Federal District Court for *patent* infringement based on the latter's addition of an electronic sensor to an existing pedal design. In turn, KSR argued that the addition of the electronic sensor was obvious, and Teleflex, Inc.'s original *patent* claim was invalid under 35 U.S.C. § 103. The Federal District Court agreed with KSR's argument that anyone with knowledge/experience in the industry would have considered it obvious to combine the previously known elements of the invention. Thus, the District Court granted summary judgement for KSR.

(*Continued*)

CLARIFICATION AND EXAMPLES (*Continued*)

Then, Teleflex appealed the case to the Court of Appeals for the Federal Circuit. The Federal Circuit Court found the lower court's analysis incomplete, applied the TSM test, and reversed the decision of the District Court. At this point, KSR appealed to the Supreme Court, arguing that the Circuit Court's test conflicted with Supreme Court precedent. The central issue in the case is the following: "In order to establish a *patent* claim's obviousness, must the courts consider the prior art, the differences between the prior art and the subject matter of the claim, and the level of ordinary skill a person must have in the subject matter of the claim, before applying secondary factors and the TSM test?" The answer is, "Indeed, yes."

The Supreme Court unanimously ruled in favor of KSR (defendant) and reversed the decision of the Federal Circuit Court.[42a] In doing so, the Supreme Court opined that the Court of Appeals *analyzed the issue in a narrow, rigid manner inconsistent with [Section 103(a)] and the precedents*. Nevertheless, the court's opinion acknowledged that a *patent* is not necessarily obvious by virtue of being a combination of two previously existing components, and that it can be helpful in such cases for a court to identify a reason that would have motivated a knowledgeable person to combine the components. However, the court also held that Federal Circuit's "TSM test" was not to be applied as a mandatory rule. The reason being that *this test for determining obviousness was too narrow, as it only considered the teachings on the specific problem the patentee was attempting to solve*. Teleflex's (plaintiff) patent was inspired by previous inventions aimed at different problems. Even though no one had combined the two pre-existing elements in the exact way Teleflex's *patent* did, the Supreme Court held that the benefit of combining the two would be obvious to any person of ordinary skill in the art. Therefore, Teleflex's (plaintiff) *patent* was obvious and invalid under 35 U.S.C. § 103.[42a]

Claims for Patent Protection—Detailed Discussion

The Concept of Claim[43]

A claim of a patent distinctly points out the scope and limits of an invention (intellectual property). The purpose of a claim is to precisely articulate what is "new" or "improved" in the invention, compared to what was already known. In this manner, *patent* claims are *similar to a survey or an abstract of a title that define the scope and limits of an owner's plot of land (physical property).* It is important to note that *patent claims define patent rights.* Thus, the *broader* the claims, the *greater will be the protection for an invention;* however, *broader* claims may be harder to get from the patent office. On the other hand, the *narrower* the claims, the thinner will be the protection for an invention, while they may be easier to receive from the patent office.

Indeed, this is *the inventor's dilemma*: whether to seek *broader claims* (difficult-to-get) for an invention and risk rejection from/delays by the *patent* office or to draft *narrower claims* (easier-to-get) and risk an easy extension/circumvention of the invention by a competing party. There are no simple answers to this question. A company selling products/services in a highly competitive industry must know that if it seeks *first-to-the-market advantage,* then that requires *being able to patent inventions faster.* On the other hand, if the company seeks *best-in-the-market advantage,* then the company must ensure that its *patents are very difficult to circumvent.* Thus, the company must engage in *strategic* IP *planning.*

In the USA, all claims are made according to the specification of 35 U.S.C. § 112. Thus, claims must be defined within one of the "statutory classes." For *utility patents,* the patentable subject matter (35 U.S.C. § 101) must fall within *machine (apparatus), process, article of manufacture,* and *composition of matter. Design patents* and *plant patents* are separate statutory classes; claims for these classes are governed under a different set of laws.

Structure of a Claim

A *patent* claim has three parts: *preamble, transitional phrase,* and *body of the claim* (Figure 3.1).

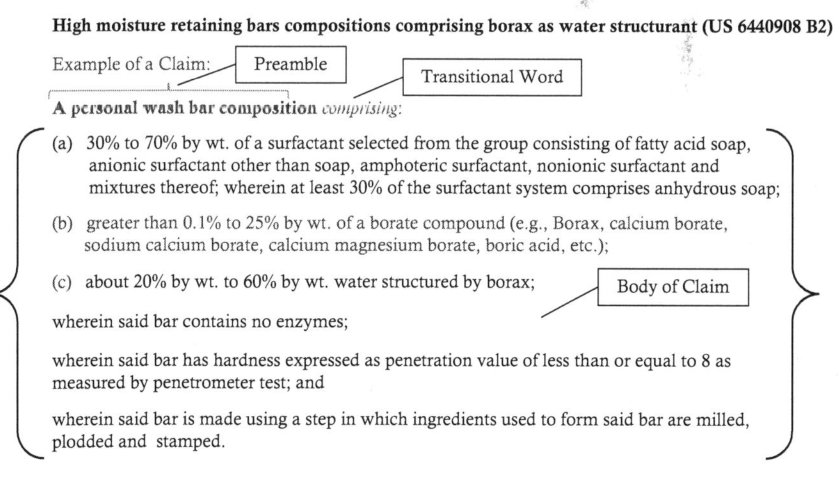

FIGURE 3.1
The structure of a claim.

Preamble. A *preamble* is an *introductory statement* that indicates *the statutory class of the invention* and *further points out what the invention is.*

The above example (US6440908)[44] is a *utility patent* (statutory class) and the invention is *an enzyme-free personal wash bar composition* (*composition of matter*).

Transitional Phrase. Generally, claims use a *transitional word* or *transitional phrase* between the *preamble* and the *body of the claim.*[45] Examples of "expansive" phrases are—"which comprises" and "comprising." The word "comprises" refers to "*including* the following elements (as listed in the *body of the claim*), but *not excluding* others." Note that the above *patent* uses the transitional word, "comprising" in its claim. The goal is to make a broad claim. On the contrary, "limiting" transitional words/phrases, such as—"including," "having," "containing," or "consisting of" are cautiously used in a claim because they convey that the invention claim is *limited only to the recited elements.* In fact, one must only use the transitional word "consisting" when the inventor is absolutely sure that additional elements would never be expected to be part of the claim.

Body of Claim. The body of a claim lists the "elements" (also referred to as "limitations") of the named invention and describes how they are interrelated, for example, how each element is related to and works with the other elements and cooperates with the whole (See Figure 3.1).

Forms of a Claim

In a *patent* (see Figure 3.1), claims can be *independent, dependent,* and *multiple dependent* (see Figure 3.2).[43–45]

A personal wash bar composition comprising: **Independent Claim**

(a) 30% to 70% by wt. of a surfactant selected from the group consisting of fatty acid soap, anionic surfactant other than soap, amphoteric surfactant, nonionic surfactant and mixtures thereof; wherein at least 30% of the surfactant system comprises anhydrous soap;

(b) 0.1% to 25% by wt. of a borate compound; **Dependent Claims**

(c) about 20% by wt. to 60% by wt. water structured by borax;

wherein said bar contains no enzymes;
wherein said bar has hardness expressed as penetration value of less than or equal to 8 as measured by penetrometer test; and
wherein said bar is made using a step in which ingredients used to form said bar are milled, plodded and stamped.

2. A composition according to claim 1, wherein surfactant system is 30-60% of bar composition.

3. A composition according to claim 1, wherein soap comprises at least 50% of surfactant system.

4. A composition according to claim 1, wherein bar has hardness expressed as penetration value less than or equal to 7.

5. A composition according to claim 1, wherein bar has hardness expressed as penetration value of 4-6.

FIGURE 3.2
The forms of a claim.

Independent Claims. An *independent* claim is one *that can stand by itself without referring to or including the limitations of other claims. Independent* claims define an operative, complete invention by themselves. The purpose of an *independent* claim is to *broadly cover all the elements of an invention.* An inventor may aim to get as many *independent* claims as allowed. However, this approach must seriously weigh the risk of rejection by the *patent* granting organization.

Dependent and Multiple Dependent Claims. *Dependent* claims specifically *refer to an independent claim and further define an invention recited in it* (See Figure 3.2). To do so, a *dependent* claim *includes all of the limitations of the claim to which it refers.* Therefore, the purpose of a *dependent* claim is to *further define the scope of the independent claim to protect the specific embodiments of that invention.* In a situation where a court finds that an *independent* claim is invalid, *dependent* claims could still be valid. Thus, *dependent* claims offer a "fall-back protection" to an invention and are very useful in infringement disputes. In addition, there are also what are known as *multiple dependent claims.* A *multiple dependent* claim *refers to more than one independent claim and further defines the invention recited in them.*

Costs Involved with Claims

Indeed, there are varying costs involved with claims in *patent* applications. In general, *utility patents* possess *at least one independent* claim and (often, but not always) *several dependent* claims. In America, the USPTO determines the filing charges for a *patent* application based on the number and forms of claims submitted. Currently, the USPTO allows three *independent* claims and 20 total claims in a *patent* application for the basic filing fee ($280). However, if a *patent* application exceeds the claims limits, the USPTO charges $420 for each additional *independent* claim (in excess of three) and $80 for each claim (beyond 20 total claims). More importantly, the USPTO charges $780 for each *multiple dependent* claim. Consequently, *patent* applications do not often use *multiple dependent* claims for this very reason. However, if the extra claims/*multiple dependent* claim offer a significant competitive advantage or intellectual capital for strategic planning (offensive/defensive), the additional fees are definitely worth it.

Different Types of Claims

There are many different types of claims, as shown below:

Method and Process Claims. Method or process claims describe a novel/improved method or process inventions that achieve a particular goal. These claims can be *independent, dependent,* or *multiple dependent.* Generally speaking, *method claims* are used to protect computer, electrical, and mechanical inventions and *process claims* are used for chemical inventions. The elements of a method or process claim are described as acts or steps.

A hypothetical example of a method claim is:

A simple 5-step procedure for operating a Wifi FPV Drone with 720P Wide-Angle HD Camera, Live Video RC Quadcopter with Altitude Hold, Gravity Sensor Function, RTF-Easy to Fly for Beginner, Compatible with VR Headset.

Note that claims are possible for methods or process steps of an invention. However, it is important to note that the steps of medical or surgical inventions that do not involve either patented devices or patented pharmaceuticals cannot be patented.

Apparatus/Machine/Device Claims. Just like the method/process claims described above, the apparatus/*machine*/device claims can also be *independent*, *dependent*, or *multiple dependent*. These claims also cover product inventions—such as, mechanical devices, electrical circuits, hydraulic *machines*, and computers with peripherals—involving collaborating parts that achieve a useful result.

A hypothetical example for an apparatus/*machine*/device claim is:

A Desk Top Home Theatre System for viewing HD movies from net streaming, comprising—a high-speed computer with a 32″ LED monitor and remote control, a state-of-the-art high-speed Internet connection and quality surround sound system for theater sound effects.

Article of Manufacture Claims. *Article of manufacture* claims are very similar to apparatus/*machine*/device claims in that they cover inventions that use a combination of collaborating elements to achieve a useful result. Unlike *machine*/apparatus, an *article of manufacture* normally involves no moving parts.

A hypothetical example of an *article of manufacture* claim is:

A specialty non-stick, dishwasher-safe, oven-safe cookware with two heat-resistant handles.

Composition of Matter Claims. A *composition of matter* claim is a product claim on the chemical composition of the compounds/substances that define the product. These claims are used in the context of chemical inventions. Figures 3.1 and 3.2 exemplify such claims.

Product-by-Process Claims. If an article or at least one element of an article is claimed by describing the process for fabricating the article or element, then that would constitute a product-by-process claim. These are also used in the context of chemical inventions.

A hypothetical example for a product-by-process claim is:

A white crystalline solid for treating prostate cancer (product), isolated from the high-pressure ethanol extraction of the bark chips of neem trees in Northern India (process).

Beauregard Claims. Beauregard claims are *computer readable media claims* named after the court decision by the same name. Such claims narrate inventions that involve a computer readable storage device—that is considered an *article of manufacture* (such as a CD/mini flash drive)—which contains commands that instruct a computer to perform a useful process.

Omnibus Claims. Omnibus claims neither "specifically point out" nor "succinctly claim" the invention. A hypothetical example of an omnibus claim is:

All the features of novelty of the golf-club bag as depicted in the Figure and described.

Some countries allow omnibus claims. However, the United States does not allow omnibus claims.

Biotechnology Claims. Biotechnology claims involve *the use of living organisms* to make products and processes. In 1987, the U.S. Commissioner of Patents and Trademarks issued a notice[23] that: "USPTO would consider non-naturally occurring, *non-human* multicellular living organisms, including animals, to be patentable subject matter within the scope of 35 U.S.C. § 101." In a 5-4 decision, in the landmark case *Diamond v. Chakrabarty*, 447 U.S. 303, 206 USPQ 193 (1980),[21] the U.S. Supreme Court ruled that while *the discovery of a natural law, natural phenomenon, abstract idea, or new mineral is not patentable* (as the subject matter is natural, abstract and/ "not made by a human"), a genetically engineered micro-organism (made by human), which is not naturally occurring (new and non-obvious), and oil-eating (useful), is *patentable* as "manufacture" and "composition of matter." However, if the living organism is a human being, *patent* protection is unconstitutional in the United States.

Jepson Claims. U.S. *patent* laws specifically allow for the protection of inventions that are *improvements* upon the work of others. According to 35 U.S.C. § 101: "Whoever invents or discovers any new and useful process, *machine*, manufacture, or *composition of matter*, or any new and *useful improvement thereof*, may obtain a *patent* therefor, subject to the conditions and requirements of this title." Thus, a Jepson claim (which is named after a 1917 court case by the same name) starts with what is *existing/known* followed by a phrase such as, wherein the *improvement* comprises, and then outlines the *new/improved* elements of the invention in support of the claim. Thus, Jepson claims first point out *prior art* and then state the *novelty/improvement* over prior art. A real-world example of Jepson claim is the Claim 9 of U.S. Patent No. 4,007,960[46]:

> In a reclining chair having a frame, a back portion, a seat portion, recliner actuator means including means for swinging said back portion between an upright and a reclined position, *the improvement comprising* elevator *means for* raising said seat and also tilting said seat forwardly to assist exit from said chair, and power actuated drive means common to both the recliner actuator means and the elevator *means for* sequentially actuating both the recliner actuator means and the elevator means and being operable sequentially in a first mode of operation to drive the recliner actuator means and operable in a second mode of operation to drive the elevator means, said power-operated drive means comprising an extensible ram, which extends in one range during said first mode of operation and in another range in said second mode of operation.

Markush Claims. Markush claims are useful in the context of chemical inventions. A Markush claim allows recitation of multiple chemical entities that are functionally equivalent to an invention. A hypothetical example for a Markush claim[47] is:

> An alkaline earth metal from the group consisting of Beryllium (Be), Magnesium (Mg), and Calcium (Ca).

Markush claims were named after Eugene Markush, the first inventor to use such claim in a *patent*. It is important to note that Markush claims are also possible with other inventions, namely—processes/*machines*/*articles of manufacture*/*compositions of matter*.

Patent Ownership

Specifics of Patent Ownership[48]

Often, people assume that *patent ownership* and *patent inventorship* are one and the same. Nothing can be farther from the truth.

Under the Patent Act (35 U. S. Code):

> One who creates an invention is its inventor, and ownership will only pass to another, including an employer, through a written assignment.

Stated differently, an employer does not automatically possess the rights to an invention created by an employee during his/her employment within the organization, if there is no written agreement in place (the exception is "hired to invent doctrine"). Thus, one can be the *inventor* in a *patent*, but not the *owner* of the *patent* rights (e.g., if the inventor assigned *patent* rights to a third-party). Similarly, a company may be the *owner* of the *patent* rights (because they were assigned the *patent* rights by the inventor), but not the *inventor*. In fact, one should note that a company is always an *assignee* and never an *inventor*. It is important to note what the USPTO says regarding the *rights* of *patent ownership*[48a]:

> *Ownership* of a *patent* gives the *patent* owner *the right to exclude others from making, using, offering for sale, selling, or importing* into the United States the invention claimed in the *patent*. 35 U.S.C. § 154 (a)(1). *Ownership* of the *patent does not* furnish the owner with *the right to make, use, offer for sale, sell, or import the claimed invention* because there may be other legal considerations precluding same (e.g., existence of another *patent* owner with a dominant *patent*, failure to obtain Food and Drug Administration (FDA) approval of the *patented invention*, an injunction by a court against making the product of the invention, or a national security related issue). A *patent* or *patent* application is assignable by an instrument in writing, and the assignment of the *patent*, or *patent* application, transfers to the assignee(s) an alienable (transferable) ownership interest in the *patent* or application, 35 U.S.C. § 261.

Therefore, the *ownership* of a *patent* must be established before filing a *patent* application to avoid future problems.

Assignment

In general, *assignment* refers to *the act of transferring the ownership of one's property* (i.e., the interest and rights to the property) *to another party*.[48a]

> According to 37 Code of Federal Regulations (CFR) § 3.1, *assignment of patent rights* is defined as: *a transfer by a party of all or part of its right, title, and interest in a patent [or] patent application....* An *assignment* of a *patent*, or *patent* application, is *the transfer to another* of a party's *entire* ownership interest or a percentage of that party's ownership interest in the *patent* or application. In order for an *assignment* to take place, *the transfer to another* must include the entirety of the bundle of rights that is associated with the ownership interest, for example, all of the bundle of rights that are inherent in the right, title, and interest in the *patent* or *patent* application.

Difference between Licensing and Assignment

It is important to note that *licensing* is "more limiting" compared to *assignment*. According to the USPTO, the following are the fundamental differences between the two[48a]:

As compared to *assignment* of *patent* rights, the *licensing of a patent transfers a bundle of rights which is less than the entire ownership interest*, e.g., *rights that may be limited as to time, geographical area, or field of use*. A *patent* license is, in effect, a *contractual agreement* that the *patent* owner will not sue the licensee for *patent* infringement if the licensee makes, uses, offers for sale, sells, or imports the claimed invention, as long as the licensee fulfills its obligations and operates within the bounds delineated by the license agreement.

An *exclusive* license may be granted by the *patent* owner to a licensee. The *exclusive* license prevents the *patent* owner (or any other party to whom the *patent* owner might wish to sell a license) from competing with the exclusive licensee, as to the geographic region, the length of time, and/or the field of use, set forth in the license agreement.

A license is *not an assignment* of the *patent*. Even if the license is an *exclusive* license, it is *not* an *assignment* of *patent* rights in the *patent* or application.

Individual and Joint Ownership—Implications[48a]

According to the USPTO, *individual* and *joint ownership* are as follows:

Individual ownership - An *individual* entity may own *the entire right, title, and interest of the patent property*. This occurs where there is *only one inventor*, and *the inventor has not assigned the patent property*. Alternatively, it occurs where all parties having ownership interest (all inventors and assignees) assign the *patent* property to one party.

Joint ownership - *Multiple parties* may *together own the entire right, title, and interest of the patent property*. This occurs when any of the following cases exist:

1. Multiple partial assignees of the patent property;
2. Multiple inventors who have not assigned their right, title, and interest; or
3. A combination of partial assignee(s) and inventor(s) who have not assigned their right, title, and interest.

Each individual inventor may only assign the interest he or she holds; thus, assignment by one joint inventor renders the assignee a *partial assignee*. A *partial assignee*, likewise, may only assign the interest it holds; thus, assignment by a partial assignee renders a subsequent assignee a partial assignee. All parties having any portion of the ownership in the *patent* property must act together as a composite entity in *patent* matters before the office.

Joint Inventorship—Implications[48b]

If an invention is made possible by the intellectual contributions of *two* or *more* inventors working together, then it constitutes *joint inventorship*. In such a scenario, *all* inventors must be named on the *patent* application, and *patent* rights will be *co-owned by the co-inventors* unless a written *patent assignment agreement* has been pre-executed by all the inventors assigning their ownership rights to another individual/company. According to U.S. law, *joint inventorship* leading to the creation of *multiple co-owners* may create the following problems[49]:

- **A Co-owner Could Independently License**. A co-owner could independently license the *patent* rights to a third-party for monetary gain, without sharing any of the gains with the other co-owners;
- **A Co-owner Could Independently Create a New Venture**. A *co-owner* could independently start a new venture and sell product(s) without infringement; and
- **All Co-owners Must Join Together to Sue an Infringer**. A *patent infringement lawsuit* against an infringer will be possible only when all *co-owners* jointly file the lawsuit.

Patent Information

According to the WIPO, "patent information" commonly refers to *the information found in patent applications and granted patents*.[50] This includes all useful information found in *patents*, such as the bibliographic data of the inventor/*patent* applicant/*patent* holder, prior art and its analysis, description of the claimed invention, and claims indicating the scope of *patent* protection. Such extensive disclosure of information in a *patent* is a result of the *patent* system's obligation *to balance* the grant of exclusive rights to an inventor *with* public disclosure of pertinent information about new/improved technology. The requirement that an inventor/*patent* applicant disclose all relevant information about the invention is crucial to other inventors, as well as technology advancement.

Indeed, *patent* information is valuable to researchers, entrepreneurs, and strategy planners, for many different reasons, such as:

- *Avoid* reinventing the wheel;
- *Assess* the latest developments in a specific field;
- *Advance* the technology to *invent* new/improved products or processes;
- *Evaluate* prior art to *determine the patentability of new/improved inventions*, in order to *apply for patent protection* nationally or internationally;

- *Monitor* technological capabilities and potential commercial interests of competitors nationally and internationally for Strengths, Weaknesses, Opportunities and Threats (SWOT) analysis;[51]
- *Avoid patent* infringement and *seek* opportunities for *licensing/buying* a *patent;*
- *Design* defensive/offensive IP strategies; and
- *Estimate* the *net worth, book value, market capitalization, and true market value* of a firm.[52]

Patent Search

"Patent search" *involves search of patent documents published by national and regional patent offices—either 18 months after the filing date of patent application or after the patent has been granted for the invention.*[53]

Through the PATENTSCOPE database,[54] the WIPO provides *free-of-charge online search access* to millions of *patent* applications filed under the Patent Cooperation Treaty (PCT) system, as well as those filed at national and regional *patent* offices, such as the European Patent Office and the USPTO. In this regard, it is also useful to note the USPTO's *seven step strategy* for *preliminary searches* of U.S. *patents* and published applications using *free online resources* of the USPTO[55a] and its bi-lateral partner the European Patent Office. The Patent and Trademark Resource Centers across America provide training on this *seven step strategy.*[55b,c]

Patent Landscaping

"Patent landscaping" is a *comprehensive, state-of-the-art study* of the *patent* and non-*patent* activities of all key players in a technology area of interest, *accompanied by insightful analysis of the entire technology landscape, business opportunities and threats, and financial implications—for the purpose of strategic planning.* Thus, *patent* landscaping involves graphical portrayal of the *patent* landscape leading *to holistic analysis of a technology area of interest* for *strategic business planning* of a company.[56b]

Patent landscaping aims to achieve the following goals[56c,d]:

- *Identify patent* and non-*patent* activities in a technology area of interest;
- *Map* the *patent* assets (e.g., granted *patents*, published and unpublished applications) of all key players *to gain a holistic perspective of the patent landscape;*
- *Correlate patent* information with non-*patent* data *to identify* business opportunities and threats, and *analyze context;*
- *Provide a clear basis for strategic business planning*—that involves exercising options, such as buying/selling/licensing of IP, innovative

> new product development, strategic new venture creation, forward and backward integration, innovative acquisitions, mergers, divestments, and strategic alliances—*in accordance with company's vision and mission, to achieve sustainable growth and competitive advantage.*

Today, research institutions and commercial firms recognize the need to protect not only individual inventions through a *patent*, but also to analyze *patent* "families" and the interactions between *patents* in a similar industry[56a] as part of *intellectual capital management*. Thus, as the forces of knowledge economy and globalization increased international competition, it has become vital for commercial firms to keep IP considerations at the front and center of *strategic business planning and innovation*, which necessitates "patent landscaping" (also known as "patent mapping").

The WIPO produces *patent* landscape reports (PLRs) for developing and under-developed nations in many areas of interest, such as *public health, food security, climate change, and the environment.*[57a] Further, the WIPO makes available many more PLRs published by international organizations, national *intellectual property* offices, non-governmental organizations, and private sector entities—either free of charge or for a fee.[57b] In addition, the WIPO also makes *useful resources* related to PLRs also accessible to all.[57b]

Finally, it is important to keep in mind the WIPO's words of advice in this regard[57c]:

> Though accessibility of *patent* information has grown as more and more *patent* offices make their *patent* documents available through online databases, certain skills are still required in order to make effective use of this information, including carrying out targeted *patent* searches and providing meaningful analysis of *patent* search results. As a result, it may be advisable to contact a *patent* information professional *for assistance where business-critical decisions are at stake.*

Patent Application Procedures

National Patent Application Procedure

One should consider the USPTO's general guidelines[58] while applying for *patents* in the USA. In this regard, the following steps will be very helpful to *patent* applicants:

- *Determine the type of intellectual property protection you need.* An applicant must first decide what is needed—a *patent, copyright, trademark,* or other types of IP protection, or any combination of these—before applying for a *patent* for an invention. The USPTO provides useful inventor and entrepreneur resources for this purpose.[59]
- *Determine if your invention is patentable.* To do this, one must do the following: Search the literature (all relevant public disclosures, domestic

and foreign *patents*, and printed publications) to establish that the invention was not publicly disclosed earlier. To do the search in the USA, a nationwide network of public, state, and academic libraries is available. However, if you are not experienced at performing *patent* searches, the USPTO recommends a list of attorneys and agents who are licensed to practice before the USPTO. In addition, if you are an inventor or small business with limited resources needing help in applying for a *patent*, the USPTO suggests that you may be eligible to receive a free attorney through Law School Clinic[60a] or Patent Pro Bono Program.[60b]

- *Determine which type of patent you need*. Since there are three types of *patents* covering different kinds of patentable subject matter, in USA, the USPTO suggests that you must decide whether you want to file a *utility patent*, *design patent*, or a *plant patent*.
- *Determine the strategy for patent filing*. Once you have determined the type of *patent* you need, the USPTO advises that you must develop a strategy for *patent* filing that fits your goals and decide whether to use professional legal services. As part of this, you must decide whether to file a provisional *patent* application[61a] first and then non-provisional *patent* later (within 12 months of filing the provisional *patent* application), or file a non-provisional *patent* application directly, whether to file a national *patent* application first and then file *patent* applications in other countries (markets) of interest, or file a PCT directly.[61b] Other considerations include the types and amounts of costs involved and time and effort that one must be prepared for.
- *Prepare and submit your initial patent application*. The USPTO suggests that inventors could use EFS-Web, the USPTO's electronic filing system for *patent* applications, to submit *utility patent* applications, provisional applications, and many other types of office correspondence to the USPTO via the Internet. Accordingly, prepare and submit your initial *patent* application for obtaining a filing date and include the correct fee.
- *Work with your patent examiner*. The USPTO wants you to understand how this process happens. If your application as submitted is incomplete, the USPTO notifies you of the deficiencies through an official letter, known as an *office action*. You will be given a specific amount of time to complete the filing (this may involve a surcharge). If the omissions are not rectified within stipulated time, your application will be either returned or dropped and the filing fee (if paid) refunded (minus handling fee).
- Next, the *patent* examiner (USPTO) will review the contents of the *patent* application to determine if the application meets all the requirements of 35 U.S.C. 111(a). If the examiner does not think your application meets the requirements, the examiner will explain the reason(s). You will have opportunities to make amendments or

argue against the examiner's objections; If you fail to respond to the examiner's requisition, within the required time, your application will be abandoned. If your application has been rejected twice, you may appeal the examiner's decision to the *Patent Trial and Appeal Board*.[62]

- *Receive your notice of allowance*. If the examiner determines that your application is in satisfactory condition and meets all the requirements, it will be approved, and you will receive a *notice of allowance*.[63] The notice of allowance will list the issue fee and may also include the publication fee that must be paid prior to the *patent* being issued.
- According to the USPTO: "Utility and reissue *patents* are issued within about four weeks after the issue fee and any required publication fee are received in the office. A *patent* number and issue date will be assigned to an application and an *Issue Notification* will be mailed after the issue fee has been paid and processed by the USPTO. The patent grant is mailed on the issue date of the *patent*. It includes any references to prior *patents*, the inventor(s') name(s), specification, and claims (to name a few). It is bound in an attractive cover and includes a gold seal and red ribbon on the cover."
- *Maintain your patent*. You must know that *maintenance fees* are required to maintain a patent in force beyond 4, 8, and 12 years after the issue date for utility and reissue *utility patents*. If the *maintenance fee* and any applicable surcharge are not paid in a timely manner, the *patent* will expire.

The national *patent* application procedures and requirements (fee structure, fee amounts, etc.) differ from one country to another, though they may share some similarities to the above procedure described for the USA. Therefore, it would be impossible to detail the *patent* application procedures for every country. On the other hand, it may be helpful to know where to look for help in the case of two major economies, China[63] and India.[64]

In addition, there are two options available to all inventors for patenting inventions in other countries: *The Paris Convention route* and *The PCT route*.[65] In the Paris Convention route, an inventor can first file a *patent* application in one's own country (which must be a member of the Paris Convention), and then file *patent* applications in other Paris Convention member countries within 12 months (12 months for utility models and 6 months for designs), claiming the priority date of the first application.

International Patent Application Procedure by the PCT Route

In this regard, the WIPO's description of PCT is instructive[66]:

> The PCT allows you to make a single *international patent application* that has the same effect as national applications filed in separate PCT states. In a nutshell, you benefit from one application, in one language paid for in one currency. The PCT assists applicants in seeking *patent* protection internationally for their inventions, helps *patent* offices with their *patent* granting decisions, and facilitates public access to a wealth of technical information relating to those inventions. By filing one international *patent* application under the PCT, applicants can simultaneously seek protection for an invention in a very large number of countries.

For a detailed discussion on this subject, see Chapter 2.

Patent Infringement Basics

What Is Patent Infringement?

According to 35 U.S.C. § 271(a), *Whoever—without authority—makes, uses, or sells any patented invention within the United States, during the term of the patent therefor, infringes the patent.*[67] Indeed, many innovators and entrepreneurs assume that a *patent* offers automatic protection against "infringement." Unfortunately, however, this is not true. In fact, when a *patent* is infringed, the *patent holder* must sue the *alleged infringer* to *stop the illegal activity (unauthorized making/using/selling of a patented invention)*, or *receive compensation for infringement.*

Since *patents* in America are governed by the U.S. Federal Law, the *patent holder* must sue the *alleged infringer* in a U.S. Federal District Court. Further, *patent holders* must take legal actions within 6 years from the date of infringement; if the lawsuit is not brought within this time period, it is time-barred, ratifying the infringement. Thus, the burden of proof lies on the *patent holder* who has to prove the infringement by a preponderance of evidence.[68]

Typically, the *alleged infringer* counters the *patent holder's* lawsuit by claiming that the original *patent* itself is invalid. *Patents* can be proven to be invalid for many reasons: (a) if the original *patent* application to the USPTO includes fraudulent information; (b) if the original *patent* is a result of anti-competitive business activities; or (c) if it can be proven later that the original *patent* did not meet the statutory requirements of novelty and non-obviousness.[68]

How Is Infringement Established?

The patent claims distinctly point out the scope and limits of an invention (intellectual property) and thereby define the patent rights. A typical *patent* contains several claims, at least one *independent* claim (often more) and several *dependent* claims. Two points are worth remembering in this regard:

First, infringing upon any one claim is considered as infringement of the whole patent. Second, infringement analysis deals *only with independent* claims.[69] This is so because *dependent* claims include every limitation of the *independent* claim on which they depend. Hence, a product/process that does not infringe upon an *independent* claim of a *patent*, cannot infringe its *dependent* claims either.

The U.S. *patent* law recognizes two types of infringement: *literal infringement* and *infringement under the doctrine of equivalents.*[70] *Literal infringement* means that the allegedly infringing product/process possesses identical correspondence to each and every element recited in an *independent* claim of the original *patent*. It is important to note that it would be *literal infringement* even when the accused product/process carries more features than the recited elements in an *independent* claim. In *infringement under the doctrine of equivalents*, the accused device *does not literally infringe the claim*. However, the accused device:

- May avoid *literal infringement* by possessing an insubstantial difference; while
- Performing substantially the same function, in substantially the same way, to achieve substantially the same result as the original invention.

Thus, the *patent holder* must prove that the *alleged infringer* has done either *literal infringement* or *infringement under the doctrine of equivalents* to halt the unauthorized making/using/selling of a *patented invention* and/or receive compensation for infringement. On the other hand, if the *alleged infringer* could *invalidate* the original *patent* of the *patent holder*, then he/she could still continue to make/use/sell the product/process. There are situations when both parties want to avoid complicated and costly lawsuits and settle matters outside the court, especially when one party is about to lose or cannot easily establish what they want in the court. Under such circumstances, the two parties may consider *licensing/cross-licensing of IP*.

In summary, *patent* infringement battles could be very complicated, lengthy, and costly. It is not easy to prove that a specific product/process infringes on the precise claims of a *patent*. However, in many cases where infringement occurs, there is sufficient evidence in terms of claims to prove it in courts to stop the unauthorized making/using/selling of a *patented invention* and/or receive compensation for infringement. Indeed, the case of *Phillips* v. *AWH Corp.*, 415 F.3d 1303 (Fed. Cir.2005) (*en banc*) is instructive on *patent* infringement, as it provides a step-by-step procedure for interpreting the claim terms.[71]

References

1. http://www.ohchr.org/EN/UDHR/Documents/UDHR_Translations/eng.pdf
2. (a) http://www.wipo.int/export/sites/www/about-ip/en/iprm/pdf/ch2.pdf; (b) http://www.wipo.int/patents/en/; (c) https://www.uspto.gov/patents-getting-started/general-information-concerning-patents; (d) http://www.makeinindia.com/policy/intellectual-property-facts; (e) http://www.china-iprhelpdesk.eu/sites/all/docs/publications/China_IPR_Guide-Guide_to_Patent_Protection_in_China_EN-2013.pdf; (f) Today, there are five regional patent offices in operation: (i) African Intellectual Property Organization (OAPI); (ii) African Regional Intellectual Property Organization (ARIPO); (iii) Eurasian Patent Organization (EAPO); (iv) European Patent Office (EPO), and (v) Patent Office of the Cooperation Council for the Arab States of the Gulf (GCC Patent Office); and (g) http://www.wipo.int/patents/en/faq_patents.html
3. (a) Racherla, U. S. "Do IPRs promote innovation" in *Innovation and IPRs in China and India: Myths, Realities and Opportunities*, Liu, K.-C.; Racherla. U. S. (Eds.,), Singapore: Springer (2016).; (b) https://law.justia.com/constitution/us/article-1/50-copyrights-and-patents.html; (c) https://www.uspto.gov/web/offices/pac/mpep/consolidated_laws.pdf; (d) https://www.uspto.gov/patent/laws-and-regulations/american-inventors-protection-act-1999/american-inventors-protection-a-2; and (e) https://www.uspto.gov/web/offices/pac/mpep/s2104.html
4. https://www.gpo.gov/fdsys/granule/USCODE-2010-title42/USCODE-2010-title42-chap23-divsnA-subchapXII-sec2181
5. http://www.wipo.int/export/sites/www/about-ip/en/iprm/pdf/ch2.pdf
6. https://www.wto.org/english/docs_e/legal_e/27-trips.pdf
7. (a) https://www.law.cornell.edu/uscode/text/35/154; (b) http://www.wipo.int/patents/en/ (c) Indian Patents Act § 53: http://www.asiaiplaw.com/article/41/1143/; (d) http://www.gnaipr.com/Acts/Indian%20Patent%20Act%20Amendment%202002.pdf; and (e) http://www.china-iprhelpdesk.eu/sites/all/docs/publications/China_IPR_Guide-Guide_to_Patent_Protection_in_China_EN-2013.pdf
8. https://www.uspto.gov/web/offices/ac/ido/oeip/taf/patdesc.htm
9. https://www.uspto.gov/patents-getting-started/general-information-concerning-patents
10. http://www.wipo.int/patents/en/faq_patents.html
11. http://www.china-iprhelpdesk.eu/content/patentsfaqs
12. script%20in%20Preparation/C-Chapters/Chapter%203/D-Reference%20Materials/China%20IPR%20Guide%20to%20Patent%20Protection%20in%20China%20(2013).pdf
13. http://www.makeinindia.com/policy/intellectual-property-facts
14. http://www.ipindia.nic.in/writereaddata/Portal/IPOGuidelinesManuals/1_28_1_manual-of-patent-office-practice_and-procedure.pdf
15. (a) https://www.uspto.gov/patents-getting-started/general-information-concerning-patents#heading-4; (b) https://www.uspto.gov/web/offices/pac/mpep/s2131.html; and (c) https://www.uspto.gov/web/offices/pac/mpep/s2158.html
16. http://www.wipo.int/patents/en/faq_patents.html

17. http://www.makeinindia.com/policy/intellectual-property-facts
18. http://www.china-iprhelpdesk.eu/content/patentsfaqs
19. (a) https://www.casebriefs.com/blog/law/patent-law/patent-law-keyed-to-adelman/patent-eligibility/gottschalk-v-benson/; (b) https://www.uspto.gov/web/offices/pac/mpep/s2106.html; and (c) https://supreme.justia.com/cases/federal/us/56/252/case.html
20. http://law.justia.com/cases/federal/appellate-courts/F2/55/854/1564867/
21. https://supreme.justia.com/cases/federal/us/447/303/case.html
22. https://www.uspto.gov/web/offices/pac/mpep/s2105.html
23. https://www.fr.com/files/Uploads/Documents/Fasse-Reiter-Baker-Industrial-Biotechnology-The-AIA-and-Its-Importance-April-2012.pdf
24. https://www.uspto.gov/web/offices/pac/mpep/s2106.html
25. https://www.gore.com/products/categories/consumer-products
26. (a) http://www.google.com/patents/US20140275122; and (b) https://www.google.ch/patents/US3929720
27. (a) https://supreme.justia.com/cases/federal/us/450/175/case.html; (b) http://www.google.com/patents/US8294; (c) http://www.google.co.in/patents/US79265; (d) http://www.google.co.in/patents/US200521; (e) https://www.google.com/patents/US101866; and (f) http://www.google.co.in/patents/US393
28. http://www.wipo.int/patents/en/faq_patents.html
29. (a) https://supreme.justia.com/cases/federal/us/409/63/case.html; (b) https://supreme.justia.com/cases/federal/us/437/584/case.html; (c) https://supreme.justia.com/cases/federal/us/450/175/case.html; (d) http://law.justia.com/cases/federal/appellate-courts/F3/33/1526/513542/; (e) https://cyber.harvard.edu/property00/patents/StateStreet.html; (f) http://law.justia.com/cases/federal/appellate-courts/F3/172/1352/599511/; (g) https://supreme.justia.com/cases/federal/us/56/62/case.html; (h) http://www.casebriefs.com/blog/law/patent-law/patent-law-keyed-to-adelman/patent-eligibility/diamond-v-diehr/; and (i) http://caselaw.findlaw.com/us-supreme-court/450/175.html
30. (a) *In re Freeman*, 573 F.2d 1237, 197 U.S.P.Q. (BNA) 464 (C.C.P.A. 1978); (b) *In re Walter*, 618 F.2d 758, 205 U.S.P.Q. (BNA) 397 (C.C.P.A. 1980); (c) *In re Abele*, 684 F.2d 902, 214 U.S.P.Q. 682 (BNA) (C.C.P.A. 1982); (d) See, Ernest Bainbridge Lipscomb III, Walker, Patents § 2:17 (3d ed. 1984) (which states: "[A] *'system' or 'method' of transacting business is not an 'art,' [i.e., process] nor does it come within any other designation of patentable subject matter* apart from the physical means of conducting the system."); and, Donald S. Chisum, Patents: A Treatise on the Law of Patentability and Infringement § 1.03[5] (1990) (which states: "*Business methods are not patentable* even though they may not be dependent upon the aesthetic, emotional, or judgmental reactions of a human"); See also, Peter D. Rosenberg, Patent Law Fundamentals, § 6.02[3] (2d ed. 1995) (which states: "Whereas an apparatus or system capable of performing a business function may comprise patentable subject matter, the law remains that *a method of doing business whether or not generated by an apparatus or system does not constitute patentable subject matter.*").
31. 35 U.S.C. § 103 (1994) specifies conditions for patentability.
32. (a) http://www.google.co.in/patents/US5333184; (b) U.S. Patent No. 5,333,184 ("the '184 patent"), teaches a method for enhancing the message record of "exchange message interface" (EMI) by adding a new field called "primary interexchange carrier" ("PIC") indicator." The '184 patent reveals many

formulas for calculating the value to be stored in the PIC indicator, which depend on the caller's and the recipient's PICs. In simplest terms, the PIC indicator may be a numerical code that identifies the recipient's PIC. Alternatively, the PIC indicator may be a Boolean value (true or false) to indicate whether the recipient's PIC is/is not the IXC that carried the call. Thus, the PIC indicator would be set to "true" if the recipient is an AT&T subscriber and "false" if not. In addition, the PIC indicator may be a Boolean value to further indicate whether or not the caller and recipient both have as their PIC, the IXC that carried the call. Accordingly, the PIC indicator for calls carried by AT&T would be set to "true" only if both caller and recipient were AT&T subscribers; and (c) Earlier, the Federal circuit upheld claims involving mathematical algorithms that "*involved the transformation or conversion of subject matter representative of or constituting physical activity or objects.*" On these grounds, the court also rejected the claims to a method of conducting an auction (business method involving a mathematical algorithm), stating that: "there is nothing physical about bids per se." See, *In re Schrader*, 22 F.3d 290. (Fed. Cir. 1994).

33. See 42 U.S.C. 2181(a).
34. McJohn, S. M. (2009). *Intellectual Property: Examples and Explanations*. New York: Aspen Publishers.
35. https://www.uspto.gov/aia_implementation/bills-112hr1249enr.pdf
36. https://supreme.justia.com/cases/federal/us/383/519/
37. (a) http://law.justia.com/cases/federal/appellate-courts/F3/185/1364/609214/; (b) http://law.justia.com/cases/federal/appellate-courts/F3/292/728/642488/
38. https://www.uspto.gov/web/offices/pac/mpep/s2141.html
39. https://www.uspto.gov/web/offices/pac/mpep/s2158.html
40. https://supreme.justia.com/cases/federal/us/383/1/case.html
41. https://www.ravellaw.com/opinions/7d6241c87f88f0cc9fb96277bf6783f5
42. (a) http://www.casebriefs.com/blog/law/intellectual-property-law/intellectual-property-keyed-to-merges/patent-law-intellectual-property-keyed-to-merges/ksr-international-co-v-teleflex-inc-2/; (b) https://www.supremecourt.gov/opinions/06pdf/04-1350.pdf; and (c) https://www.oyez.org/cases/2006/04-1350
43. (a) http://thelimitedmonopoly.com/wp-content/uploads/2015/05/200904LimitedMonopoly-StakingYourPatentClaimsPartI.pdf; and (b) http://www.patent-education.com/images/200905_Limited_Monopoly_-_Staking_Your_Patent_Claims_Part_II.pdf
44. https://www.google.com/patents/US6440908
45. http://www.bios.net/daisy/patentlens/2352.html
46. (a) https://www.google.com/patents/US4007960; and (b) http://www.ipwatchdog.com/2012/03/29/patent-claim-drafting-improvements-and-jepson-claims/id=23580/
47. http://www.jafarilawgroup.com/markush-claims-use-one-use-one/
48. (a) https://www.uspto.gov/web/offices/pac/mpep/s301.html; and (b) https://www.neustel.com/patent-ownership-basics/
49. http://www.ipeg.com/avoid-jointly-owned-intellectual-property/
50. http://www.wipo.int/patents/en/faq_patents.html#info
51. https://www.businessballs.com/strategy-innovation/swot-analysis-19/
52. See chapter 1
53. See reference 50 for further details.

54. (a) http://www.wipo.int/patentscope/en/; and (b) http://www.wipo.int/edocs/pubdocs/en/patents/434/wipo_pub_l434_03.pdf
55. (a) https://www.uspto.gov/patents-application-process/search-patents; (b) https://www.uspto.gov/learning-and-resources/support-centers/patent-and-trademark-resource-centers-ptrc/resources/seven; and (c) https://www.uspto.gov/sites/default/files/documents/7_Step_US_Patent_Search_Strategy_Guide_2015_rev.pdf
56. (a) http://www.sciencemag.org/careers/2001/10/patent-mapping; (b) http://www.caduceustechnology.com/index.php?option=com_content&view=article&id=100&Itemid=94; (c) http://www.iam-media.com/Intelligence/IP-Value/2011/Legal-perspectives-North-America/United-StatesThe-whys-and-hows-of-patent-landscaping; and (d) https://www.ipcheckups.com/blog/patent-landscape-analysis-overview/
57. (a) http://www.wipo.int/patentscope/en/programs/patent_landscapes/; (b) http://www.wipo.int/patentscope/en/programs/patent_landscapes/plrdb.html; and (c) http://www.wipo.int/patents/en/faq_patents.html
58. https://www.uspto.gov/patents-getting-started/patent-process-overview
59. (a) https://www.uspto.gov/learning-and-resources/inventors-entrepreneurs-resources; and (b) https://www.uspto.gov/learning-and-resources/support-centers/inventors-assistance-center-iac
60. (a) https://www.uspto.gov/learning-and-resources/ip-policy/public-information-about-practitioners/law-school-clinic-1; and (b) https://www.uspto.gov/patents-getting-started/using-legal-services/pro-bono/patent-pro-bono-program
61. (a) https://www.uspto.gov/patents-getting-started/patent-basics/types-patent-applications/provisional-application-patent; and (b) https://www.uspto.gov/patents-getting-started/patent-basics/types-patent-applications/nonprovisional-utility-patent
62. https://www.uspto.gov/patents-application-process/patent-trial-and-appeal-board-0.html
63. http://www.china-iprhelpdesk.eu/sites/all/docs/publications/China_IPR_Guide-Guide_to_Patent_Protection_in_China_EN-2013.pdf
64. https://ipindiaonline.gov.in/epatentfiling/UsefullDownloads/Comprehensive%20efiling%20manual.pdf
65. See chapter 2 for a detailed discussion on this subject.
66. http://www.wipo.int/pct/en/
67. https://www.law.cornell.edu/uscode/text/35/271
68. http://smallbusiness.findlaw.com/intellectual-property/patent-infringement-and-litigation.html
69. https://www.evans-dixon.com/DMS-%232761740-v2-Website_Article_Patent_Infringement_Basics.pdf
70. http://www.intellectual.com/content/patent-infringement
71. https://www.casebriefs.com/blog/law/patent-law/patent-law-keyed-to-adelman/infringement/phillips-v-awh-corporation/

4

Copyrights

The Congress shall have Power...To promote the *Progress of Science and useful Arts,* by securing *for limited Times to Authors* and *Inventors the exclusive Right to their respective Writings* and *Discoveries.*

The United States Constitution, Title 1, Section 8

Concept of Copyright

Generally speaking, a *copyright* is a *form of intellectual property right* grounded in the Constitution and granted by law for original works of authorship fixed in a tangible medium of expression. A *Copyright* covers both *published* and *unpublished* works.

According to the European Union Copyright Office:[1]

> *Copyright* is a legal term describing the rights given to creators for their literary and artistic works. The purpose of a *copyright* is to allow creators to gain economic rewards for their efforts and so encourage future creativity and the development of new material which benefits us all. *Copyright* material is usually the result of creative skill and significant labor and investment, and without protection, it would often be very easy for others to exploit material without paying the creator. *Copyright* protection *is automatic as soon as there is a record in any form of the work created.*

According to the United States Copyright Office (USCO)[2]:

> *Copyright* is a form of protection provided by the laws of the United States to the authors of original works of authorship that are fixed in a tangible form of expression. An original work of authorship is a work that is independently created by a human author and possesses at least some minimal degree of creativity. A work is "fixed" when it is captured (either by or under the authority of an author) in a sufficiently permanent medium such that the work can be perceived, reproduced, or communicated for more than a short time. *Copyright* protection in the United States exists *automatically from the moment the original work of authorship is fixed.*

The U.S. Copyright Law is contained in the **Title 17 of the United States Code** (17 U.S.C.).[3a] Title 17 U.S.C. contains the *Copyright Act of 1976* and subsequent amendments in Chapters 1–8 and 10–12. Chapter 9 includes the *Semiconductor Chip Protection Act* of 1984, as amended; and chapter 13 includes the *Vessel Hull Design Protection Act*, as amended.[3b] The USCO is responsible for registering intellectual property claims under all three acts.

Subject Matter of Copyrights—Copyrightable Categories

The following overview applies to *all copyrightable works of authorship* in the United States of America (USA).[4a]

According to 17 U.S.C. § 102(a)[4b]:

> The subject matter of *copyright* protection includes *original works of authorship fixed in any tangible medium of expression, now known or later developed,* from which they can be perceived, reproduced, or otherwise communicated, *either directly or with the aid of a machine or device.*

The following are the *copyrightable categories of authorship*:

1. Literary works;
2. Musical works, including any accompanying words;
3. Dramatic works, including any accompanying music;
4. Pantomimes and choreographic works;
5. Pictorial, graphic, and sculptural works;
6. Motion pictures and other audio-visual works;
7. Sound recordings; and
8. Architectural works.

For the purpose of registration, one must understand that *these categories are interpreted very broadly* in U.S. Copyright Law. Thus, for example, computer programs and certain data bases are registrable as "literary works;" maps and technical drawings can be registered as "pictorial, graphic, and sculptural works." In addition, *these categories are not mutually exclusive*. Thus, for example, a play could fall into two categories—"literary" and "dramatic works." Furthermore, a work *that may not* fall into any of the above eight categories may still qualify for a *copyright*, so long as *it is an original work of authorship fixed in tangible form*. Therefore, one must recognize that the above categories are intentionally broad to accommodate future possibilities presently not imaginable. However, according to 17 U.S.C. § 102(b),[5a] *copyright* protection for *original works of authorship* does not extend to *ideas, procedures, processes, systems, methods of operation, concepts, principles, or discoveries,* regardless how they are described, explained, illustrated, or embodied in such work.[5b]

Subject Matter of Copyrights—Details of Copyrightable Categories

Literary Works

In 17 U.S.C. § 101,[6] "Literary works" are defined as:

> Works, *other than audio-visual works,* expressed in words, numbers, or other verbal or numerical symbols or indicia, regardless of the nature of the material objects, such as books, periodicals, manuscripts, phonorecords, film, tapes, disks, or cards, in which they are embodied.

Thus, a "literary work" is a *non-dramatic textual work* that explains, describes, or narrates a particular subject, theme, or idea through a narrative, descriptive, or explanatory text, rather than dialog or dramatic action. Hence, *non-dramatic literary works are generally intended to be read* and *not intended to be performed before an audience.* Consequently, *textual works*—that are created for performance before an audience or sound recording, motion picture, or *audio-visual* purpose—are not considered as *non-dramatic literary works;* instead, they are considered as the *works of the performing arts.*[7] The *works of the performing arts* include—musical works including any accompanying words, sound recordings, *dramatic works* including any accompanying music, pantomimes and choreographic works, motion pictures, and other *audio-visual* works, each of which will be taken up separately below.

"Literary works," include *many categories of original works* such as—fiction, non-fiction, poetry, directories, catalogs, reference works, advertising copies, compilations of information, computer programs, and data bases, to mention a few.

It is important to note that, in the above list, a computer program, also known as "computer software," pushes the boundary of what the literature scholars consider as "literary work." Nevertheless, *computer program is a copyrightable work* whether in *source code, object code,* or *other form.* In fact, the U.S. Congress amended the Copyright Act of 1980 to accommodate *computer program* as *copyrightable* work, defining it as: "a set of statements/instructions to be used directly/indirectly in a computer to bring about a certain result," 17 U.S.C. § 101.[6] In addition, the U.S. Congress introduced some limitations on the *copyright* of a computer program to define its scope, 17 U.S.C. § 117.[8]

Musical Works (Including any accompanying words)

"Musical works" (also known as *musical compositions*) are *original works of authorship consisting of music and any accompanying words.*[9a] "Musical works" comprise of all genres (such as *country music, pop music, jazz, heavy metal, hip-hop, a cappella,* etc.,) and sub-genres (such as *country blues, pop country, rap,* etc.).

According to the United States Copyright Office[9b]:

> The *author(s) of a musical composition is (are)* generally *composer,* and *lyricist,* if they are different. A musical composition may be in the form of a *notated copy* (for example, sheet music) or in the form of a *phonorecord* (for example, cassette tape, LP, or CD).

A *dramatic musical work is a musical work created for use* in a motion picture or a *dramatic work,* including musical plays and operas. On the contrary, a *non-dramatic musical work is not a musical work created for use* in a motion picture or a *dramatic work,* such as a ballad made for distribution exclusively on an album, or an advertising jingle intended solely for performance on the radio.

Sound Recordings

According to 17 U.S.C. § 101[10a]:

> *Sound recordings* are works that result from the fixation of a series of musical, spoken, or other sounds, but not including sounds accompanying a motion picture or other *audio-visual work,* regardless of the nature of the material objects, such as disks, tapes, or other phonorecords, in which they are embodied.

Generally, a *sound recording* is a *recorded performance* of *another work.* A *sound recording* must be fixed. In other words, the sounds must be captured in a medium from which they can be perceived, reproduced, or otherwise communicated. The author of a *sound recording* may choose to fix the sounds in a digital track, disk, tape, or other format.[10b] Generally, *sound recordings* are recordings of music, such as—a song, vocal performance, musical performance, a *literary work* (e.g., an audiobook), a digital file of a performance, or similar recordings. However, *sound recordings* may also include subject matter, such as—bird calls, a hurricane wind, a forest fire, a waterfall, or a classroom lecture by a student—which require creative human effort.

It is important to point out that a *musical composition* and a *sound recording of that musical composition by an artist* are distinctly separate works. For example, *the song* "Just beat it, beat it" and *a sound recording* of it by Michael Jackson are distinctly different works. Thus, while the song itself (i.e., the music and lyrics) is a *musical work,* the audio recording of that song by Michael Jackson is a *sound recording.* Therefore, the *copyright* in a *musical work* covers only the *musical composition* (music and lyrics, if any), but not the *sound recording* of the composition. Similarly, the *copyright* in a *sound recording* covers *only* the recording, *but not* the underlying *musical work*, for example, music, lyrics, words, or other content embodied in that recording.

The U.S. Copyright Law explicitly states that the definition of a *sound recording* "does not include" the "sounds accompanying a motion picture or other audio-visual work," 17 U.S.C. § 101 (see above). Thus, when an applicant prepares to register the sounds in a motion picture or other *audio-visual work*, he/she must state "sounds," "soundtrack," or "sounds accompanying a motion picture/audio-visual work," rather than "sound recording."

Motion Pictures, and Other Audio-Visual Works

According to 17 U.S.C. § 101[11a,b]:

> *Audio-visual works* are works that consist of a series of related images which are intrinsically intended *to be shown by the use of machines or devices* such as projectors, viewers, or electronic equipment, together with accompanying sounds, if any, regardless of the nature of the material objects, such as films or tapes, *in which the works are embodied.*
>
> *Motion pictures* are *audio-visual works* consisting of a series of related images which, when shown in succession, impart an impression of motion, together with accompanying sounds, if any.
>
> *Motion picture exhibition facility* means a movie theater, screening room, or other venue that is being used primarily for the exhibition of a copyrighted motion picture, if such exhibition is open to the public or is made to an assembled group of viewers outside of a normal circle of a family and its social acquaintances.

Audio-visual works broadly cover any work that *includes series of related visual images*, "whether or not moving, and with or without sounds," as long as a machine or device is needed to view the related series of images. Thus, *audio-visual works* include *motion pictures*, and *much more* (such as, video games, PowerPoint slide presentations, online *audio-visual works* e.g., smartphone and tablet applications, online courses and tutorials, website content). On the other hand, a circus act, a stage play, or a puppet show—*are not audio-visual works*, as *they do not require a machine or device to view/hear them.*

Thus, when an applicant deposits the material to assert claim in a "motion picture," the registration specialist will try to establish *if the images impart some kind of motion*. The definition of "motion pictures" *excludes*: (a) unauthorized fixations of live performances/telecasts; (b) live telecasts not fixed simultaneously with their transmission; and (c) filmstrips/slide sets incapable of conveying an impression of motion, though they may consist a series of images intended to be shown in succession.

Dramatic Works (Including any accompanying music)

According to the USCO Compendium of U.S. Copyright Office Practices (3rd Edition)[12]:

> *Dramatic works* are not defined in the U.S. Copyright Law. However, for the sake of *copyright* registration, one should know that *dramatic works* are *compositions in prose or verse that portray a story intended to be performed for an audience*. Generally, a *dramatic work* represents the action as it occurs rather than simply narrating or describing the action.

It is important to note the differences between dramatic works, *non-dramatic literary works*, and *non-dramatic musical works*.

Dramatic works do not simply narrate or describe an action; instead they represent the action as it occurs and *are intended to be performed for an audience*. Examples of *dramatic works* include stage plays, musicals, or operas. Also, *dramatic works* may/may not include music. On the contrary, *non-dramatic literary works* (such as novels) are very different from *dramatic works* (such as plays, musicals, or operas) in that they—unlike *dramatic works—are not intended to be performed for an audience*. Similarly, *non-dramatic musical works* (such as a symphony) are also different from *dramatic works with accompanying music* (such as an opera or a musical), as the *former do not portray a story*, whereas the *latter do convey a story*.

Pantomimes and Choreographic Works[13]

According to USCO Compendium of U.S. Copyright Office Practices (3rd Edition)[12a]:

> *Pantomime* is, the art of imitating, presenting, or acting out situations, characters, or events through the use of physical gestures and bodily movements. Long before Congress extended federal *copyright* protection to *pantomimes*, the Supreme Court recognized that *a silent performance* is worthy of *copyright* protection if it qualifies as a *dramatic work*.
>
> Similarly, *choreography* is the composition and arrangement of a related series of dance movements and patterns organized into a coherent whole.

The U.S. Copyright Act recognizes *choreography* as a distinct category of *copyrightable authorship*, 17 U.S.C. § 102(a)(4). Here, it should also be noted that *choreography* and *dance* are not synonymous.[12b] A *dance* is a static and kinetic succession of bodily movements in certain rhythmic and spatial relationships and in relation to time and space. Thus, *choreography* is a distinct subset of dance that encompasses certain types of compositional dances. According to the legislative history for the 1976 Copyright Act, *choreographic works* do not include *social dance steps* and *simple routines*. Examples of this category include—a mime, ballet, a choreographed dance in a Bollywood movie, a synchronized swimming display, or a gymnastics floor exercise routine.

Pictorial, Graphic, and Sculptural Works

According to USCO Compendium of U.S. Copyright Office Practices (3rd Edition)[12a]:

> *Pictorial, graphic, and sculptural works* include two-dimensional and three-dimensional works of fine, graphic, and applied art, photographs, prints and art reproductions, maps, globes, charts, diagrams, models, and technical drawings, including architectural plans. Such works shall include works of *artistic craftsmanship* insofar as their form, but not their mechanical or utilitarian aspects, are concerned; the design of a useful article, as defined in this section, shall be considered a *pictorial, graphic,* or *sculptural work* only if, and only to the extent that, such design incorporates *pictorial, graphic,* or sculptural features that can be identified separately from, and are capable of existing independently of, the utilitarian aspects of the article 17 U.S.C. § 101.

The details of the *copyrightable* works in this category of are as follows:

The class of *pictorial works* include—paintings, drawings, photographs, prints, art reproductions, maps, technical drawings, diagrams, applied art (i.e., two-dimensional pictorial artwork applied to a useful article), artistic crafts (e.g., textiles, table service patterns, wall plaques), online or digital artwork (e.g., computer-aided artwork, digital imaging, pixel art), and similar works.

The class of *graphical works* also include—drawings, prints, art reproductions, maps, technical drawings, diagrams, applied art (i.e., two-dimensional graphic artwork applied to a useful article), artistic crafts (e.g., textiles, table service patterns, wall plaques), online or digital artwork (e.g., computer-aided artwork, digital imaging, pixel art), and similar works.

Finally, the class of *sculptural works* include—sculptures, globes, models, applied art (i.e., three-dimensional artwork applied to a useful article), artistic crafts (e.g., jewelry, glassware, toys, dolls, stuffed toy animals, models), and similar works.

For the full scope of exclusive rights afforded by *pictorial, graphic,* and *sculptural works,* see 17 U.S.C. § 113.[12b]

Architectural Works

According to 17 U.S.C. § 101:

> An *architectural work* is the design of a building as embodied in any tangible medium of expression, including a building, architectural plans, or drawings. The work includes the overall form as well as the arrangement

> and composition of spaces and elements in the design, but does not include individual standard features.

The U.S. Copyright Office registers as architectural works designs for structures that can be *inhabited by humans* or are otherwise *intended for human occupancy.*

Some examples include—houses, office buildings, churches, museums, gazebos, basketball arenas, football stadiums, and garden pavilions.

Subject Matter of Copyrights—Compilations and Derivative Works

According to 17 U.S.C. § 103,[14a,b] *copyright* protection extends to *compilations* and *derivative works.* However, protection for a work employing *pre-existing material in which copyright subsists* DOES NOT extend to *any part of the work that uses the material unlawfully.*

The *copyright* in a compilation or *derivative* work extends *only to the material contributed by the author of such work,* as distinguished from the pre-existing material employed in the work and DOES NOT imply any exclusive right in the pre-existing material. The *copyright* in such work is independent of, and DOES NOT affect or enlarge the scope, duration, ownership, or subsistence of, any *copyright* protection in the pre-existing material.

Compilations

A "compilation" is defined as:

> A work formed by the collection and assembling of pre-existing materials or of data that are selected, coordinated, or arranged in such a way that the resulting work as a whole constitutes an original work of authorship, 17 U.S.C. § 101.

The term "compilation" also includes *collective works.*

Examples include—a *telephone directory* of a city, that creatively compiles "useful" phone numbers of several kinds, including businesses, doctor's offices, neighborhood clinics, hospitals, police precincts, government offices etc.; a *handbook* of chemistry & physics that compiles vast amounts of "useful" data, laws, tables, equations, data conversion methods, units for measurement etc.; a *compendium* of the "best" poems before 1950; a *collection* of the "most popular" *sound recordings* of Michael Jackson's era; a *photographic compilation of* the "world's most influential" political leaders; a *website* of an online business that compiles "attractive" photos, graphics, and text for all products available for purchase; an *academic monograph* containing "original research" articles by many authors on the sustainability of earth's resources; and a *newspaper* with multiple "breaking stories" written by different investigative journalists. In the examples listed above, *original work of authorship* is shown in deciding *useful, best, most popular, world's most influential, attractive, original, and breaking stories.*

Derivative Works

A "derivative work" is defined as:

> A *work* based upon one or more pre-existing works, such as a translation, musical arrangement, dramatization, fictionalization, motion picture version, *sound recording,* art reproduction, abridgment, condensation, or any other form in which a work may be recast, transformed, or adapted. A *work* consisting of editorial revisions, annotations, elaborations, or other modifications, *which, as a whole, represent an original work of authorship,* 17 U.S.C. § 101.

Derivative works are a *subset* of the subject matter categories listed in § 102 (of the Copyright Act), rather than a *separate* and *distinct* category of work. In other words, the new material that the author contributed to the *derivative work* must fall within one or more of the eight categories detailed above. Examples of *derivative works* include—a *translation* of a novel written in one specific language (let us say English) into other languages (such as, Chinese, French, German, Hindi etc.); a *musical arrangement* of a pre-existing *musical work*; a *motion picture* such as, "Where Eagles Dare" *based on the novel* by the same name, written by Alister MacLean; a *dramatization* of the play, such as Macbeth by Shakespeare; an *abridgment* of a novel, such as *A Tale of Two Cities* by Charles Dickens; a *fictionalization* of the Bible; a *revision* of a previously published book; a *sculpture* based on a drawing; a *drawing* based on a photograph; and a *photograph* based on a sculpture, etc.

Subject Matter of Copyrights—National Origin

This is described in detail in 17 U.S.C. § 104.[15a] This subject matter comprises of:

1. **Unpublished works[15b]**

   As per this title, the works specified under 17 U.S.C. § 102 and 17 U.S.C. § 103, *are subject to protection, even when unpublished,* without regard to the nationality or domicile of the author;

2. **Published works**[15b]

According to this title, *published works* specified under 17 U.S.C. § 102 and 17 U.S.C. § 103, when published, *are subject to protection under this title* if:

1. On the date of first publication, one or more of the authors is a national or domiciliary of the United States, or is a national, domiciliary, or sovereign authority of a treaty party, or is a stateless person, wherever that person may be domiciled;
2. The work is first published in the United States or in a foreign nation that, on the date of first publication, is a treaty party;
3. The work is a *sound recording* that was first fixed in a treaty party;
4. The work is a *pictorial, graphic,* or *sculptural work* that is incorporated in a building or other structure, or an architectural work that is embodied in a building and the building or structure is located in the United States or a treaty party;
5. The work is first published by the United Nations or any of its specialized agencies, or by the Organization of American States; or
6. The work comes within the scope of a Presidential proclamation. Whenever the President finds that a particular foreign nation extends, to works by authors who are nationals or domiciliaries of the United States or to works that are first published in the United States, *copyright* protection on substantially the same basis as that on which the foreign nation extends protection to works of its own nationals and domiciliaries and works first published in that nation, the President may by proclamation extend protection under this title to works of which one or more of the authors is, on the date of first publication, a national, domiciliary, or sovereign authority of that nation, or which was first published in that nation. The President may revise, suspend, or revoke any such proclamation or impose any conditions or limitations on protection under a proclamation.

   For purposes of paragraph (2), a work that is published in the United States or a treaty party within 30 days after publication in a foreign nation that is not a treaty party shall be considered to be first published in the United States or such treaty party, *as the case may be.*

3. **Effect of Berne Convention**[15b]

No right or interest in a work eligible for protection under this title may be claimed by virtue of, or in reliance upon, the provisions of the Berne Convention, or the adherence of the United States thereto. Any rights in a work eligible for protection under this title that derive from this title, other than Federal or State statutes, or the common law, shall not be expanded or reduced by virtue of, or in reliance upon, the provisions of the Berne Convention, or the adherence of the United States thereto.

4. **Effect of Phonograms Treaties**[15b]

   Notwithstanding the provisions of sub-section (b), no works other than *sound recordings* shall be eligible for protection under this title solely by virtue of the adherence of the United States to the Geneva Phonograms Convention or the WIPO Performances and Phonograms Treaty.

Subject Matter of Copyright—U.S. Government Works

According to 17 U.S.C. § 105[15c]:

> *The works of the U.S. Government*—defined in the law as works prepared by an officer or employee of the U.S. Government as part of that person's official duties—*are not copyrightable.*

works (other than the edicts of the U.S. government) prepared by officers of any government (state, local, or foreign governments, except the U.S. Government), are subject to registration, if they are otherwise *copyrightable.*[15d]

Eligibility and Ineligibility of Works for Copyright Protection

Based on 17 U.S.C. § 102 (a),[4b] it is clear that:

A work is *eligible* for *copyright* protection if it is—

1. Original;
2. Work of authorship;
3. Fixed in a tangible medium of expression, now known or later developed, from which they can be perceived, reproduced, or otherwise communicated, either directly or with the aid of a machine or device; and
4. Provided the work falls *within* the categories of *copyrightable* subject matter.

On the other hand, a work may be *ineligible* for *copyright* protection if it:

1. Fails to meet any of the *above substantive requirements* (even if the work falls is in a category of *copyrightable* subject matter); and,
2. Falls *within* the *non-copyrightable categories of works* (described later in this chapter).

CLARIFICATION AND EXAMPLES

Thus, based on 17 U.S.C. § 102 (a), a work must meet three substantive requirements to be eligible for *copyright* protection. Let us now look at the details.

1. **Original.** Thus, the *first condition for eligibility of copyright protection* is that the work be "original." For this, a work must pass two specific tests: (i) be *independently created by the author* (not copied, acquired, or stolen); and, (ii) possess a *minimal degree of creativity.*

   Examples: The *copyright* in a work initially belongs to the author(s) who created that work. An *acquired painting, a copied architectural design, or a piece of stolen art*, are *ineligible* for *copyright* protection by anyone (even if the artists who created them are unknown), because they fail the first test: "not independently created" by those that wish to acquire a *copyright*; an *unorganized random list of all the phone numbers* in a city of businesses, doctor's offices, neighborhood clinics, hospitals, police precincts, government offices, etc., will be *ineligible* for *copyright* protection because it fails the second test: "does not possess a minimal degree of creativity." On the other hand, even the telephone listings of a city are *copyrightable* if a "minimal degree of creativity" can be demonstrated. See, *Key Publications v. Chinatown Today Publications Enterprises, 945 F. 2d 509 (2d Cir. 1991).*[16]

   "Originality" *does not imply* "novelty" or "uniqueness." Thus, a work that may contain "non-original subject matter" and resembles another work, may still be *eligible* for *copyright* protection *as long as it shows a minimal degree of creativity.* The example below illustrates the point. Suppose, an author wrote a book on the *Natural Wonders of the World* and got a *copyright* for it. Let us assume that five years later, a different author *independently* wrote another book on "the same subject (without copying)," not knowing that his subject matter closely resembles that of the first book. Indeed, the second book

(*Continued*)

CLARIFICATION AND EXAMPLES (*Continued*)

(original work of authorship) will be also *eligible* for *copyright* protection though it lacks "novelty" or "uniqueness." It must also be noted that "originality" *does not imply* "quality," "artistic merit," or "ingenuity."

2. **Work of authorship.** The *second condition for eligibility of copyright protection* is that the work be: (i) be a *work of authorship* and that (ii) *it owes its origin to a human being.* This is very broadly interpreted. Thus, a "work of authorship" is a *product of creative expression by a human author* within the *copyrightable* categories of subject matter. Hence, it must be clear that works produced solely by nature/plants/animals/mechanical means will be *ineligible for copyright protection.*

According to the USCO Copyright Practices[17]:

> In order to be entitled to *copyright registration*, a work must be *the product of human authorship.*
>
> Works *produced by mechanical processes* or random selection *without any contribution by a human author are not registrable.* Thus, a linoleum floor covering featuring a multicolored pebble design which was produced by a mechanical process in unrepeatable, random patterns *is not registrable.*
>
> Similarly, a work owing its form to *the forces of nature* and *lacking human authorship is not registrable*; thus, a piece of driftwood, even if polished and mounted, *is not registrable.*

Mechanical/photomechanical reproductions are *ineligible* for *copyright* protection.[17] Thus, a *microfilm reproduction* of "The Tragedy of Julius Caesar" by William Shakespeare (public domain textual matter), or a *color photocopy* of the Mona Lisa painting by Leonardo da Vinci (public domain painting) *are not registrable.*

Similarly, a *random art painting* accidentally produced by two cats fighting on a spread canvas on the floor when paint bottles are open is *ineligible* for *copyright* protection even if it looks fantastic because the artwork is *not created by a human.* Another such *ineligible example* for *copyright* protection is a *practice drama act* mechanically captured on a tape by a security camera fixed in the lobby of a hotel, because it is not a creative expression by a human author.

(*Continued*)

CLARIFICATION AND EXAMPLES (*Continued*)

3. **Fixed in a tangible medium of expression.** The *third condition of eligibility for copyright protection* is that a work "be fixed in a tangible medium of expression by or under the authority of the author."[17]

   According to 17 U.S.C. § 101:

   > A *work is fixed* in a tangible medium of expression when its embodiment in a *copy* or *phonorecord*, by or under the authority of the author, is sufficiently permanent or stable to permit it to be perceived, reproduced, or otherwise communicated for a period of more than transitory duration. A work consisting of sounds, images, or both, that are being transmitted, is fixed for purposes of this title if a fixation of the work is being made simultaneously with its transmission.[18]

   The terms, *copies* and *phonorecords* are also defined very broadly.

   > *Copies* are material objects, other than *phonorecords*, in which a work is *fixed by any method now known or later developed, and from which the work can be perceived, reproduced, or otherwise communicated, either directly or with the aid of a machine or device.* 17 U.S.C. § 101.
   >
   > *Phonorecords* are "material objects in which sounds—other than those accompanying a motion picture or other *audiovisual work*—are *fixed by any method now known or later developed, and from which the sounds can be perceived, reproduced, or otherwise communicated, either directly or with the aid of a machine or device.*" This term "includes the material object in which the sounds are first fixed." 17 U.S.C. § 101.

   Thus, it is important to note that a work becomes *eligible* for *copyright* protection *only after it is fixed in a tangible medium of expression by/under the authority of the author.*

   Accordingly, a *speech* or *musical work* will be eligible for *copyright* protection only after the speaker *fixes his speech* and a musician *fixes the song*. A speaker can fix his speech by writing it down on paper or making a voice recording of it. Similarly, a musician can fix his *song* through sheet music or *sound recording*. A *drama* may be embodied in manuscript, typescript, printed copy, on a video-recording, such as a videocassette, or another form of copy, or in a phonorecord. The fixation of a drama may be made simultaneously with its transmission or live performance.

Non-Copyrightable Works[19a,b]

The U.S. Copyright Law [17 U.S.C. § 102(b)] declares that the following categories of works are *non-copyrightable*:

1. Ideas, procedures, methods, systems, processes, concepts, principles, or discoveries;
2. Works that are not fixed in a tangible form (such as a choreographic work that has not been notated or recorded or an improvisational speech that has not been written down);
3. Titles, names, short phrases, and slogans;
4. Familiar symbols or designs;
5. Mere variations of typographic ornamentation, lettering, or coloring; and
6. Mere listings of ingredients or contents.

CLARIFICATION AND EXAMPLES

We shall clarify *non-copyrightable* works below with examples.

1. **Ideas, procedures, methods, systems, concepts, principles, or discoveries.** As mentioned above, 17 U.S.C. § 102(b) *expressly excludes copyright protection for*: "any idea, procedure, process, system, method of operation, concept, principle, or discovery, regardless of the form in which it is described, explained, illustrated, or embodied." However, the USCO may register a literary, graphic, or artistic description, explanation, or illustration of an idea, procedure, process, system, or method of operation, *provided that the work contains a sufficient amount of original authorship.*

   The following examples clarify the point.

   Suppose, Peter observed that a herd of deer regularly visit the backyard of his country home on every Sunday evening (discovery) and decided to record one such episode (idea/concept). One day, he actually created a motion picture (method) of it, using a high-speed camera, coupled with lighting and sound effects (system). Peter used his family members to control lighting and sound effects, while he creatively filmed the event and added a voice commentary later (procedure). Then, Peter got *copyright* protection for his motion picture (in the next section,

(*Continued*)

CLARIFICATION AND EXAMPLES (*Continued*)

we will describe the exclusive rights received by a *copyright* owner). Peter's *copyright* protection extends *only to the original expression* and *not to the underlying discovery, idea/concept, method, system, or procedure.* Thus, Peter's *copyright does not prohibit* his neighbor, Paul, from making *another recording of such an episode* on another Sunday, using exactly the same method, system, and procedure, and even receiving a *copyright* for his original work.

Let us look at a different example. Suppose, Susan *discovered* a new animal in a jungle. Then, Susan elaborately described the animal for the first time (original, work of authorship) in a handwritten note in her own diary (fixation). The pages of the diary—in which Susan recorded a descriptive summary of the animal—*are copyrightable.* However, while Susan receives a *copyright* for *creative expression,* the *discovery* itself is not protected. What it means is that Susan's *copyright does not prevent* anyone to find the same animal in the jungle and record a new description of the same animal in his/her own words and claim another *copyright* for that work.

2. **Works not fixed in tangible form.** As already pointed out, an original work of authorship, *must be fixed in tangible form* in order to receive a *copyright.* The following example illustrates the point. Suppose, Jack made an extemporaneous speech on "leadership" *without* writing it down, or making a *sound recording* of it, while his friend Jill was actively listening. In fact, Jill was so impressed with the speech that she went ahead and gave the same speech on stage in her college. Unfortunately, Jack could not have anything to stop Jill from making the speech he composed, because his work *was not fixed in tangible form and hence has no copyright.*

   Similarly, *choreographic works that have not been notated or recorded* have no *copyright* protection, because *they have not been fixed.*

3. **Names, titles, short phrases, and slogans**. Names, titles, short phrases or expressions, and slogans are not *copyrightable,* even if they are novel, distinctive, or lend themselves to a play of words. Similarly, a mere listing of ingredients or contents is not *copyrightable.*[19b] See, 37 C.F.R. 202.1(a).[19c] To be protected by a *copyright,* a work must contain *sufficient amount of creativity.* Names, titles, and other short phrases are simply too minimal to meet these requirements.

   Examples of names, titles, or short phrases that do not contain a *sufficient amount of creativity* to support a claim in *copyright*

(*Continued*)

CLARIFICATION AND EXAMPLES (*Continued*)

include—the *name of an individual* (including pseudonyms, pen names, or stage names); the *name of a business* or *organization*; the *name of a product* or *service*; the *name of a band* or *performing group*; the *name of a character*; a *domain name* or URL; the *title* or *subtitle* of a work—such as a book, a song, or a *pictorial, graphic,* or *sculptural work*; *catchwords* or *catchphrases*; and *mottos, slogans,* or other *short expressions.* Some of them are the subject of trademarks.

4. **Familiar symbols or designs.** Familiar *symbols* or *designs* are not *copyrightable.* See 37 C.F.R. 202.1(a).[19c] *Typeface* is not *copyrightable, nor* is the *design, format,* or *layout* of books and other printed material. Familiar symbols or designs, or a simple combination of a few familiar symbols or designs, are *uncopyrightable* and cannot be registered with the office. However, a *work of authorship* that incorporates one or more familiar symbols or designs into a larger design may be registered if the work as a whole shows a *sufficient amount of creativity.*

   Examples include—*letters, punctuation, or symbols* on a keyboard; *abbreviations;* musical *notation;* numbers and mathematical and currency *symbols;* arrows and other *directional* or *navigational symbols; common symbols* and shapes, such as a spade, club, heart, diamond, star, yin yang; *common patterns,* such as standard chevron, polka dot, checkerboard, or hounds tooth; well-known and commonly used symbols that contain a minimal amount of expression or are in the public domain, such as the *peace symbol, gender symbols, or simple emoticons;* industry *designs,* such as the caduceus, barber pole, food labelling symbols, or hazard warning symbols; familiar *religious symbol;* and common *architecture molding.*

5. **Mere variations of typographic ornamentation, lettering, or coloring.** A typeface is a set of letters, numbers, or other characters with repeating design elements that is used in composing text or other combinations of characters, including *calligraphy.* Generally, typeface, fonts, and lettering are *building blocks of expression* that are used to create *works of authorship.* The office cannot register a claim to *copyright* in typeface or mere variations of typographic ornamentation or lettering, regardless of whether the typeface is commonly used or unique. Mere variations of *typographic ornamentation, lettering,* or *coloring,* are *not copyrightable.* See 37 C.F.R. 202.1(a).[19c]

(*Continued*)

CLARIFICATION AND EXAMPLES (*Continued*)

6. **Mere listings of ingredients or contents.** A mere listing of ingredients or contents, or a simple set of directions, is *uncopyrightable.* As a result, the office cannot register recipes consisting of a set of ingredients and a process for preparing a dish. In contrast, a recipe *that creatively explains* or *depicts how or why to perform a particular activity may be copyrightable.* A registration for a recipe may cover the written description or explanation of a process that appears in the work, as well as any photographs or illustrations that are owned by the applicant. However, the registration will not cover the list of ingredients that appear in each recipe, the underlying process for making the dish, or the resulting dish itself. The registration will also not cover the activities described in the work that are procedures, processes, or methods of operation, which *are not copyrightable.*[20]

The Exclusive Rights of a Copyright Owner

According to 17 U.S.C. § 106:[21] *copyright* owner possesses *six* "exclusive rights" *to do* and *to authorize* the following, subject to the limitations on exclusive rights, as specified under 17 U.S.C. § 107–122[22]:

1. **Reproduction right**. To *reproduce the copyrighted work in copies/phonorecords;*
2. **Adaptation right**. To *prepare derivative works based on the copyrighted work;*
3. **Public distribution right**. To *distribute copies/phonorecords of the copyrighted work to the public* by sale or other transfer of ownership, or by rental, lease, or lending;
4. **Public performance right**. To *perform the copyrighted work publicly* in the case of *literary, musical, dramatic,* and *choreographic works, pantomimes,* and motion pictures and other audio-visual works;
5. **Public display right**. To *display the copyrighted work publicly* in the case of *literary, musical, dramatic,* and *choreographic works, pantomimes,* and *pictorial, graphic,* or *sculptural works,* including the individual images of a motion picture or other *audio-visual work;* and
6. **Public digital audio transmission right**. To *perform the copyrighted work publicly* by means of a digital audio transmission in the case of *sound recordings.*

A U.S. *copyright automatically protects the original work internationally in all the states that are signatories to the Berne Convention.* However, there may be nuances to the particular national laws applicable in these states (see WIPO Lex).[23]

Limitations on Exclusive Rights: Fair Use Doctrine

The *unauthorized use of a copyrighted work* is *copyright infringement* and may subject the infringer to civil and criminal penalties under federal law. Thus, in general, the duplication, dissemination, or appropriation of a *copyrighted* work, without the permission of the original author, is prohibited. Similarly, the public display or performance of *copyrighted* works is also restricted. *Fair use* is *the right to use a copyrighted work without permission of the copyright owner* (under certain conditions), to prevent a rigid application of the law that would otherwise stifle creativity. It allows one to use and creatively build upon previous works without unfairly depriving the rights of prior *copyright* owners to control and benefit from their works.

The U.S. Copyright Law is similar to the *copyright* law in many countries that are party to the Berne Convention, except that in the USA the *fair use* is more broadly defined. According to 17 U.S.C. § 107:[24] Notwithstanding the provisions of 17 U.S.C. § 106 and § 106 A, *the fair use of a copyrighted work*—including such use by reproduction in copies or phonorecords or by any other means specified by that section, for purposes such as criticism, comment, news reporting, teaching (including multiple copies for classroom use), scholarship, or research—*is not an infringement of copyright*. In order to determine whether the use of a work is *fair use*, the following factors must be considered:

- The *purpose and character of the use*, including whether such use is of a commercial nature or is for nonprofit educational purposes;
- The *nature of the copyrighted work*;
- The *amount and substantiality of the portion used* in relation to the *copyrighted* work as a whole; and
- The *effect of the use upon the potential market for or value of the copyrighted work*.

The fact that a work is unpublished shall not itself bar a finding of *fair use* if such finding is made upon consideration of all the above factors. A detailed discussion of this subject can be found elsewhere.[25]

The Scope of Exclusive Rights and Copyright Infringement

According to 17 U.S.C. § 501(a)[26a]:

> *Anyone who violates any of the exclusive rights* of the *copyright* owner as provided by sections 17 U.S.C. § 106 through 17 U.S.C. § 122 or of the author as provided in section 106A(a), or who imports copies or phonorecords into the United States in violation of section 602, *is an infringer of the copyright or right of the author,* as the case may be.

Simply stated, according to the USCO[26b]:

> *Copyright infringement occurs* when a *copyrighted* work is reproduced, distributed, performed, publicly displayed, or made into a *derivative work* without the permission of the *copyright* owner.

To establish *copyright infringement,* the plaintiff must prove[27]: (1) the *ownership of a valid copyright,* and (2) *actionable copying* by the defendant of constituent elements of the work that are original. Thus, to establish the first condition (namely, ownership of a valid *copyright*), the plaintiff must prove that the work was *independently created* by the author, and that it *possesses at least a minimal degree of creativity.* In addition, the plaintiff must demonstrate compliance with the statutory formalities, for example, obtaining *copyright registration* in a timely manner. To establish the second condition (namely, actionable copying), a plaintiff must prove that the defendant did *actually copy* from the plaintiff's work (called as *factual copying*), and that the *works are sufficiently similar* when compared, to establish appropriation.

Reproduction Right

According to 17 U.S.C. § 106(1)[21]:

> A *copyright* owner has the exclusive right to do/authorize the *reproduction of the copyrighted work* in *copies* or *phonorecords.*[28]

For example, the author of a book (*literary work*) may make photocopies or pdf copies of his book, while the author of a musical composition (*musical work*) may make copies of her work in sheet music or phonorecords, such as in cassette tape, flash drive, or CD. The same is applicable to the authors of all other kinds of works described in 4.2.1.

Anyone who makes *unauthorized* copies or phonorecords of a *copyrighted* work, potentially *infringes.* A *copyright* is, indeed, a *strict liability.* Therefore, courts require the plaintiff to minimally prove *volition or causation* by the defendant. Infringement does not depend on the *knowledge of the statute, intent to infringe,* or *motive for monetary gain.*[29]

"Independent creation of a work" that may resemble a *copyrighted* work is *not infringement,* however close that resemblance might seem.

On the other hand, either *direct copying,* namely, actual copying a *copyrighted* work, or *indirect copying,* namely, actual copying of an intermediary work that copied the original *copyrighted* work—both constitute *infringement.*

Therefore, to establish infringement, the *copyright* owner (plaintiff) must prove that the accused (defendant) "actually copied" the *copyrighted* work either *directly* or *indirectly.* This is why it is sometimes called as: "factual copying."

Further, to establish infringement, the plaintiff must prove that the defendant *actually copied the copyrighted subject matter* and *not some non-original expression, ideas, facts, or other functional elements, which are not copyright protected.* In addition, the plaintiff must prove that the defendant's work *resembles* the original *copyrighted* work *sufficiently closely.*

Adaptation Right

According to 17 U.S.C. § 106(2)[21]:

> A copyright owner has the exclusive right *to prepare the derivative works*[30] *by himself/herself* or *to authorize someone to do so based on his/her copyrighted work.*

Thus, the author of an English novel can: (a) *translate the novel into other languages* (such as, French, German, etc.); (b) *recast the novel into a motion picture;* (c) *adapt the novel into a Broadway play;* (d) *abridge* the novel; (e) *fictionalize* the novel; and, (f) *revise* the novel. Similarly, an artist can (a) *sculpt* based on one's own drawing and (b) *take a photograph* of one's own painting. These are just few examples.

A *derivative work* receives a *separate copyright.* To be *copyrightable,* a *derivative work* must bring in *new original copyrightable authorship,* while incorporating some or all of a pre-existing work. Two points must be made here about the *copyright* of the *derivative work*: firstly, it does not apply to any portion of the *derivative work* in which material was used unlawfully, 17 U.S.C. § 103(a);[14a] and secondly, it applies only to the material contributed by the author of the *derivative work,* 17 U.S.C. § 103(b).[14a]

Infringement occurs when someone recasts/transforms/adapts a *copyrighted* original work into a *derivative work—without obtaining prior authorization from the author of the original work.* The "**fair use**" **doctrine** is an exception to this. Examples of this include:

1. *Works of scholarship* that include portions of a *copyrighted* work that are "absolutely critical" to the point being made, to make creative advancements;

2. *Book reviews* that use the original language of a *copyrighted* work as necessary, to make authentic observations or comments; and
3. *Parodies* that use *copyrighted* works to make fun of, criticize, or otherwise comment on those original works.

Public Distribution Right

According to 17 U.S.C. § 106(3)[21]:

> A *copyright* owner has the exclusive right to *do/authorize* the distribution of copies/phonorecords of his/her *copyrighted* work to the public by sale/other transfer of ownership or rental/leasing/lending.

Thus, a writer has the right to sell/rent/lease/lend copies of her text book. However, it would be *infringement* if a book seller sells, rents, leases, or lends *unauthorized* photocopies of the author's textbook. Similarly, a musician has the right to distribute copies of his CD containing most popular hits through sale, rental, leasing, or lending. On the contrary, it would be *infringement* if any other person sells, rents, leases, or lends *unauthorized* copies of the same CDs.

Nevertheless, the public distribution right is limited by the "**first sale doctrine**," which states that:

> Notwithstanding the provisions of 17 U.S.C. § 106(3), the owner of a particular copy or phonorecord lawfully made under this title, or any person authorized by such owner, is entitled, without the authority of the *copyright* owner, to sell or otherwise dispose of the possession of that copy or phonorecord, 17 U.S.C. § 109(a).[32]

Thus, after the first sale or distribution of a copy, *the copyright holder has no control over what happens next to that copy of the work.* Stated differently, the purchaser of a legal copy of *copyrighted* work can decide what to do next with the copy he/she owns, as long as the *copyright* owner's *exclusive rights are not infringed.*

Two examples illustrate this doctrine.[33] Suppose, a book has been lawfully purchased at a book store (the first sale). The *copyright holder* will have no control over what can happen to the copy of that book. Thus, the *new owner* of the copy of that book may choose to *sell, rent, destroy, or even freely give it away without anyone's permission.*[31] Similarly, a DVD rental store may lawfully purchase movie DVDs (the first sale) and rent them to the public, *without paying any royalties to the copyright holder of the movie.*

Fortunately, this need not be of major concern as: (a) the "first sale doctrine" is *narrow* and applies *only* to the specific copy/copies; and (b) *no rights* to the underlying work are granted to the owner. As a result, all that the owner can do is to sell, rent, destroy, or even freely give away the particular copy that he legally owns. In fact, the new owner *cannot* reproduce, adapt, publish, or perform the work *without the authorization of the author*. For example, the "first sale doctrine" does not allow a music CD owner to copy some of the songs from that CD onto a different CD, add personal commentary, and sell the new CD. Furthermore, only the owner of a work can benefit from the "first sale doctrine," not any person who merely possesses the copy without owning it. For example, while the DVD rental store that owns movie DVDs can rent the DVDs to the public, the people renting the DVDs cannot go ahead and rent the copies they temporarily possess anymore.

Public Performance Right

According to 17 U.S.C. § 106(4)[21]:

> A *copyright* owner has the exclusive right to do/authorize the *performance of the copyrighted work publicly* in the case of *literary, musical, dramatic,* and *choreographic works, pantomimes,* and motion pictures and other *audio-visual works.*

To perform or display a work *publicly* means—

1. *To perform* [the work] or display it at a place open to the public or *at any place where a substantial number of persons outside of a normal circle of a family and its social acquaintances is gathered*; or
2. *To transmit or otherwise communicate a performance* or display *of the work* to a place specified by clause (1) or to the public, by means of any device or process, whether the members of the public capable of receiving the performance or display receive it in the same place or in separate places and at the same time or at different times, 17 U.S.C. § 101.[34]

A *private performance* of a *copyrighted* work is a performance to oneself, one's family, or a small group of friends. This is allowed. On the other hand, *public performance* of a *copyrighted* work is a performance at any *public gathering venue*. This constitutes *infringement*. This includes any *copyrighted* work, such as—*literary, musical, dramatic,* and *choreographic works, pantomimes,* motion pictures, and other *audio-visual works.*

Some examples may be helpful. Parents reciting a *copyrighted* poem at home to their children every night, is perfectly allowed. On the other hand, an unauthorized recitation of the same poem in a Broadway play would be *infringement*. Similarly, if one sings Whitney Houston's "Greatest Love of All," to one's family, with guitar and key board accompaniments, it poses no problems. However, an unauthorized rendering of the same song on stage at a public auditorium as part of a music show would be *infringement*. Further, if a drama party enacts a *copyrighted* play at home to a small group of family members, it would be allowed. However, if the same drama party performs the *copyrighted* play on stage at a public auditorium, it would be *infringement*. Similarly, a radio or television broadcast, Internet performances, and music-on-hold performances are all considered public performances. Finally, a restaurant owner/bar owner playing music CDs/movie DVDs to the business customers would constitute *infringement*, even if the *copyrighted* works were lawfully purchased.

This is so because the rights to publicly perform the work still reside with the *copyright* owner. Therefore, *the only way to avoid the infringement problem is to obtain licenses to the copyrighted works prior to the public performance from license granting organizations*. The organizations that *grant performance licenses* on behalf of songwriters, composers, and music publishers, are—the *American Society of Composers, Authors and Publishers* (ASCAP), *Broadcast Music Incorporated* (BMI), and the *Society of European Stage Authors & Composers* (SESAC). In this regard, it is important to know that ASCAP, BMI, and SESAC possess their own repertoires of music. Therefore, a business must decide which songs of their interest are in the repertoires of ASCAP, BMI, and SESAC by searching their individual websites. Then, that business must purchase a license from the organization. Thus, ASCAP, BMI, and SESAC collect licensing fees for performed works and distribute the revenues as royalties to the corresponding *copyright* owners.

According to 17 U.S.C. § 504, the *copyright* owner or author has a choice of recovering: (a) actual damages and any additional profits of the defendant or (b) statutory damages. For example, in a case of "innocent infringement," the range is $200 to $30,000 per work, while in a case of "willful infringement," damages can be increased to up to $150,000 per work.[35a] Therefore, all businesses must clearly understand the benefits of licensing, prior to public performance of any *copyrighted* work. Indeed, through compliance with the *copyright* law, businesses can avoid disastrous *copyright* damages caused by infringement.[35b]

Public Display Right

According to 17 U.S.C. § 106(5)[21]:

> A *copyright* owner has the exclusive right to do/authorize the *display of the copyrighted work publicly* in the case of *literary, musical, dramatic,* and *choreographic works, pantomimes,* and motion pictures, and other *audio-visual works.*

According to 17 U.S.C. § 101:

> To display a work means to show a copy of it, either directly or by means of a film, slide, television image, or any other device or process or, in the case of a motion picture or other audiovisual work, to show individual images non-sequentially.

Further, we discussed what "to display publicly" means, in the above section.

Thus, the *public display right* is similar to the *public performance right*, except that the former is applicable to public "display" rather than a "performance." Both of these rights apply to *musical works* (the lyrics, composition, and arrangement), but not to *sound recordings* (a particular version of a *musical work*).[36a]

A public display of a *copyrighted* work means to show a visual copy of it to others. This right covers *individual images* (stills) from a film, *reproductions* of paintings and drawings, *sheet music* from *musical works*, or *photos* from other performance pieces. When one uses slides, films, or television, the condition for public display will be satisfied, directly or indirectly. It is also important to distinguish a public display from a private display. Just as in the case of public performance vs. private performance, it would be a private display if one shows *copyrighted* work to a small circle of family and friends. On the contrary, it would be a public display when the showing is done at a public venue for anyone and everyone.

Some examples may be helpful in this regard. If a person collects his favorite photographs from different *copyrighted* works and displays them at home to a small circle of family members, there would be no issues. On the contrary, if that person takes the same set of photographs and displays them at a public venue such as a *museum*, that would constitute infringement. Similarly, if a person selects paintings and drawings from different magazines and makes a slide show to a small group of friends at home, that would be fine. However, if the person displays the same slide show on a *television program* to the general public, that would constitute infringement. Further, if the same slide show is *transmitted on the Internet* to a large group of people, that would also constitute infringement.

According to some experts[36b] in the field:

> *Copyright* holders and some courts have urged the "RAM copy" doctrine[36c] in an attempt to extend *copyright* to computer networks. But the doctrine threatens to create significant problems as more and more works are used in digital form and has the potential to give *copyright* owners excessive control over the use of their works. Reinvigorating the public display right—the tool Congress intended and designed to address this particular technological advance—would provide a more balanced mechanism for protecting *copyright* owners' incentives to create and exploit works in a digitally networked world without unduly diminishing access to those works.

Public Digital Audio Transmission Right

According to 17 U.S.C. § 106(6)[21]:

> A *copyright* owner has the exclusive right to do/authorize the performance of the *copyrighted* work publicly by means of a digital audio transmission in the case of *sound recordings*.

The *Digital Performance Right in Sound Recordings Act* (DPRSRA)[37a,b] enacted in 1995 grants the owners of a *copyright* in sound recordings an exclusive right "to perform the copyrighted work publicly by means of a digital audio transmission." The DPRSRA was enacted for two reasons: (a) the Copyright Act of 1976 *did not specify the performance right for sound recordings* and (b) the legitimate fears of businesses *that digital technology would suppress the sales of physical records.*[37c] The performance right for *sound recordings* under the DPRSRA *is limited only to digital audio transmission,* and not as expansive as the performance right for other types of *copyrighted* works.[37d] Hence, performers criticize the DPRSRA's comparative inequity, pointing out that composers have a much wider performance right than the performers.[37e]

A hypothetical example could be helpful in this regard. When you hear John Doe's *sound recording* of "I Love You Baby" on a radio in the USA, Jane Russell—the composer of "I Love You Baby"—is compensated through BMI, while John Doe receives nothing for this public performance. This is because terrestrial broadcasters in the USA *are exempt from paying a public performance right for sound recordings.*

On the other hand, if you hear the same performance on Sirius XM, via a webcast, or on a cable music station—even if the webcast is done by a terrestrial broadcasting station—both Jane Russell and John Doe are compensated. The *Digital Millennium Copyright Act* of 1998, addressed some of the issues in DPRSRA.[37f]

Copyright Registration

Some people may question: "Why do I need *copyright registration,* when *copyright* protection is automatic?" According to the European Union Copyright Office:[38] "Many choose to register their works because they wish to have the facts of their *copyright* on the public record and have a certificate of registration. Registered works may be eligible for statutory damages and attorney's fees in successful litigation."

In order to understand *copyright registration* in Europe, one must learn about the history of formality-free *copyright* protection.[39] Before 1886, every nation had its own rules for *copyright* recognition, and so the authors of original works complied with the formalities on a country-by-country basis. The Berne Convention of 1886, however, changed that situation by introducing a rule that required authors belonging to the signatory nations to comply only with the formalities of the country in which the work originated. In 1908, the Berlin revision of the Berne Convention introduced the current rule of formality-free *copyright* protection. Under this new rule, the enjoyment and the exercise of a *copyright* shall not be subject to any formality. See, the Paris Act of 1971, Article 5.[40]

In accordance with the principles established by the Berne Convention, several member states developed voluntary *copyright registration systems.*

The goal of a *copyright registration system* is to create a clear and effective means to establish authorship and/or ownership of rights, to facilitate the exercise of *copyright* and related rights. As a result, today, the national registration and recordation systems maintain valuable information on creativity, from the legal and economic angles. Thus, a *copyright registration system* can provide not only the certificates of registration, but also the certified copies of registry documents—that *offer key information about a work, including authorship/ownership through a documented chain of transfer. Copyright* registration also *enables public access to the creative content of the copyrighted works,* for which authorization from the *copyright* owner is not required. In addition, the information contained in the national registries can also serve as *a source of national statistics on creativity and culture.* Finally, national registries may serve as repositories of cultural heritage, as they hold historical collections of national creativity.

In recent years, rapid development of digital technologies, coupled with increasing numbers of creators (frequently unidentified) and breath-taking flow of content, highlight the role of accurate and reliable data regarding ownership and the clear need for documentation and recordation. Thus, rapidly evolving digital economy has raised new issues concerning the registration of *copyright* and related rights. Some experts argue that *copyright registration* can play a significant role in addressing the challenges associated with creative content in the case of "orphan works," for which the *copyright* owner cannot be identified/located.

According to USCO:[41a]

> The U.S. Copyright Office does not issue *copyrights,* but instead simply registers claims to a *copyright.* See 17 U.S.C. § 408(a) (stating that: "the owner of *copyright* or of any exclusive right in the work may obtain registration of the *copyright* claim" by submitting an appropriate application, filing fee, and deposit to the Copyright Office). The *copyright* in a work of authorship created or first published after January 1, 1978 is protected from the moment it is created, provided that the work is original and is fixed in a tangible medium of expression, 17 U.S.C. §§ 102(a), 408(a). In other words, the *copyright* in a work of original authorship exists regardless of whether the work has been submitted for registration or whether the office has issued a certificate of registration for that work. See 17 U.S.C. § 408(a) ("registration is not a condition of *copyright* protection").

Thus, an application for *copyright registration* in the USA contains three elements: a *completed application form,* a *non-refundable filing fee,* and a *non-returnable deposit* (a copy or copies of the work being registered and "deposited" with the Copyright Office).[41b] When the USCO issues a *registration certificate, it assigns the date it received all required elements in acceptable form as the effective date of registration,* irrespective of the time it took to process the application and mail the certificate of registration. See elsewhere for *full details on copyright registration* in the USA.[41b]

Though it is not required for *copyright* protection, USCO recommends *copyright registration* for the following reasons[41c]:

1. It *serves as a public record* comprising of key facts relating to the authorship and ownership of the claimed work. Further, it *provides related information* about the work, such as title, year of creation, date of publication (if any), and the type of authorship that the work contains (e.g., photographs, text, *sound recordings*);
2. It (or refusal to register) is *a prerequisite for filing a lawsuit on copyright infringement* involving a U.S. work. 17 U.S.C. § 411(a);[41d]
3. In a *copyright* infringement lawsuit, *statutory damages and/or attorney's fees can be claimed* when *copyright registration* is done prior to the infringement or within three months of the first publication of the work, 17 U.S.C. § 412(c), § 504, § 505;
4. It works as a *prima facie evidence of the validity of the copyright* and *the facts stated in the certificate of registration*, provided the work is registered before or within 5 years of the first publication of the work;
5. It *provides essential information to prospective licensees*—such as name and address of the *copyright* owner;
6. A document recorded with USCO may *provide constructive notice of the facts stated therein*, provided the document identifies the work of authorship and only when that work has been registered, 17 U.S.C. § 205(c)(1) & (2);
7. The deposition of copy(ies) submitted with an application for *copyright registration* of a published work *may satisfy the mandatory deposition requirement*, only if the applicant submitted the best edition of the work, 17 U.S.C. § 407, § 408(b);
8. It is a prerequisite for receiving the *full benefits of a pre-registration*[1] that has been issued by USCO, 17 U.S.C. § 408(f)(3);
9. The U.S. Customs and Border Protection *can seize foreign pirated copies of a copyright owner's work*, provided that the work had been registered with USCO and the registration certificate had been recorded with the U.S. Customs and Border Protection; and
10. To receive royalties under compulsory license, a *copyright owner must be identified in the registration or other public records*[2] of USCO, 17 U.S.C. § 115(c)(1).[41d]

---

[1] *Pre-registration* is a special service that is intended for specific types of works that are likely to be infringed before they are completed or before they are released for commercial distribution, such as feature films.

[2] *Registration* and *recordation* are very different procedures: claims to *copyright* are registered, while documents related to *copyright* claims are recorded, such as agreements to transfer or grant a mortgage in *copyrights*.

Copyright Notice

Copyright notice is *a statement placed on copies* or *phonorecords of a work* by which a *copyright* owner informs the public of his/her claim of ownership.

Accordingly, a *copyright notice* comprises of three elements (1, 2, and 3 shown below) that appear as a single continuous statement:

1. The symbol © for *copies* or the symbol ℗ for *phonorecords;*
2. The *year of first publication* of the work; and
3. The name of the *copyright owner.*

Two examples for *copyright notice* are: © 2017 ABC Press and ℗ 2018 Taylor Ritchie.

The use of a *copyright notice* is the responsibility of the copyright owner and does not require *permission from,* or *registration with,* the Copyright Office.

- For all works first published *before* March 1, 1989, *copyright notice was required,* subject to few exceptions. During this period, a work lost *copyright* protection in the United States if *copyright notice* was omitted or a mistake was made in using the *copyright notice;*
- For all works published *on/after* March 1, 1989, as well as unpublished works, and foreign works, *copyright notice is optional;*

Although *copyright notice is optional* for unpublished works, foreign works, or works published on/after March 1, 1989, USCO strongly encourages *copyright* owners to use a *copyright notice* for the following reasons:[42]

- This *notifies* potential users that a *copyright* is claimed in the work;
- With a published work in a *copyright* infringement action, a *copyright notice* may thwart a defendant *from attempting to limit his/her liability for damages or injunctive relief* based on an innocent infringement defense;
- For parties seeking permission to use the work, a *copyright notice* is helpful, as it identifies the *copyright* owner at the time the work was first published;
- It *identifies the year of first publication,* which is helpful in cases of an *anonymous work,* a *pseudonymous work,* or *a work made for hire*—to determine the term of *copyright* protection. Note: These works will be described in the next section; and
- It may prevent the work from becoming an *orphan work* by identifying the *copyright* owner and/or specifying the term of the *copyright.* See 17 U.S.C. §§ 401(d), 402(d), 405(b), 406(a), and 504(c)(2). Full details can be found elsewhere.[42]

Copyrights in Different Contexts—Authorship and Ownership

It is always important to know *the authorship and ownership of* a *copyright* in different contexts, such as—*individual works, joint works, works made for hire, derivative works, compilations, collective works, anonymous works,* and *pseudonymous works.*

The *copyright* in a work of authorship initially belongs to *the author* or *co-authors* of that work, unless and until the *copyright* has been assigned by author(s) to another party in a signed, written agreement or by operation of law, 17 U.S.C. § 201(a) and 17 U.S.C. § 204(a).[43a] However, *if the author does not own the copyright in the work any longer,* the applicant must provide a brief statement to explain *how the claimant obtained ownership of the copyright,* 17 U.S.C. § 409(5). For guidance on these matters, see Chapter 600 § 619 and § 620.[43b]

1. **Individual works.** Any individual who—by himself/herself—creates a work that meets the conditions of originality, work of authorship, and fixation in a tangible medium (assuming that he/she was not authorized by anyone to do so), is recognized as *the author of the individual work, from the point of fixation.* It is important to note that the *copyright ownership* in individual works *initially belongs to the author of the original work.*

   Thus, Alister MacLean is the author of the individual work *Where Eagles Dare.* Similarly, Alfred Tennyson is the author of the classic individual work of poems, *In Memoriam.* However, in certain contexts, *where many people may be involved in producing a work, a single person may still be deemed as the author of the work.* In such cases, the courts will depend on key factors such as:

   a. Who originated the work;
   b. Who contributed to the creative expression of the work;
   c. Who controlled the production of the work; and
   d. Who is regarded as the author by others associated with the work/third parties.

2. **Joint works.** The U.S. Copyright Law defines a "joint work" as:

   > A work prepared by two or more authors with the intention that their contributions be merged into inseparable or interdependent parts of a unitary whole, 17 U.S.C. § 101.[44a]

   According to the USCO, a work of authorship is considered a "joint work:"

If the authors collaborated with each other, or if each of the authors prepared his or her contribution with the knowledge and intention that it would be merged with the contributions of other authors as inseparable or interdependent parts of a unitary whole.[44b]

Indeed, the *intention* that the individual parts will be absorbed or combined into an integrated work, *at the time the work was created, is key.*[44c]

Thus, a contribution to a joint work is considered: (a) "inseparable" if the work contains a single form of authorship, such as a novel or painting and (b) "interdependent" if the work contains multiple forms of authorship, such as motion picture, opera, or the music and lyrics of a song.[44d] According to the USCO, the applicant—and not the USCO—must determine whether a work qualifies as a joint work. This determination must be based on existing facts at the time the work was created. With regards to the ownership of a *copyright*, the authors of a joint work *jointly own the copyright* in each other's contributions and each author owns an undivided interest in the *copyright* for the work as a whole, 17 U.S.C. § 201(a).[44e]

3. **Works made for hire.** The term "work made for hire" is defined in 17 U.S.C. § 101. This definition applies to *works created on/after* January 1, 1978.

The U.S. Copyright Law defines "work made for hire" as[45]:

a. A work *prepared by an employee within the scope of his or her employment*; or
b. A work *that is specially ordered or commissioned, provided that the parties expressly agree in a written instrument signed by them* that the work shall be considered a "work made for hire," and provided that the work is *specially ordered or commissioned* for use as:
   - A contribution to a collective work;
   - A part of a motion picture or other *audio-visual work;*
   - A translation;
   - A compilation;
   - A test;
   - An answer material for a test;
   - An atlas;
   - An instructional text, which is defined as *a literary, pictorial, or graphic work prepared for publication and with the purpose of use in systematic instructional activities*; or
   - A supplementary work, which is defined as *a work prepared for publication as a secondary adjunct to a work by another author for the purpose of introducing, concluding, illustrating, explaining, revising, commenting upon, or assisting in the use of the other work,*

> *such as forewords, afterwords, pictorial illustrations, maps, charts, tables, editorial notes, musical arrangements, answer material for tests, bibliographies, appendixes, and indexes,"* 17 U.S.C. § 101.

In this regard, some examples will be helpful.

A *"specially ordered or commissioned work"* will be *a work made for hire* if it meets the following *four* requirements[45]:

- The work *must fall within the categories of works* listed above in the definition;
- There *must be an express written agreement* between the party ordering or commissioning the work and the individual(s) who create the work;.
- The agreement *must state that the work shall be considered a work made for hire*; and
- The agreement *must be signed by both the parties.*

No work *that fails to satisfy all four requirements,* will be a *work made for hire.*

CLARIFICATION AND EXAMPLES

1. *Work created by an employee within the scope of his/her employment*

   Example-1. Peter is a manufacturing engineer at the ABC Consumer Products company. His job is to scale up the process of making a hair and body cleansing formulation and develop the specifications for large scale manufacture of a New Hair & Body Wash. Having completed the job successfully, Peter created an *instruction manual* for the manufacture of New Hair & Body Wash, containing easy-to-follow step-wise procedures. Indeed, Peter (an employee) is the creator of this original work, but he did this work within the scope of his employment with the ABC Consumer Products company (employer). Therefore, this work is a *work made for hire* and ABC Consumer Products company should be named as the *author of the work* in the application for *copyright* registration.

   Example-2. Janet is a creative sound engineer and a full-time employee of EMI Music company, which produces *pop music recordings.* Janet conducts all of her professional work using the company's sound mixing equipment at the company studios. Janet brings several budding artists and creates pop music recordings of new artists. Clearly, these *sound recordings* by Janet (employee) are within the scope of her employment with EMI Music company (employer). Hence, this work

(*Continued*)

CLARIFICATION AND EXAMPLES (*Continued*)

is a *work made for hire* and EMI Music company would be named as *the author of the work* in the application for *copyright* registration.

2. *Work created by an employee outside the scope of his/her employment*

   Example-1. Paul is a Pulitzer prize winning staff-writer for the daily newspaper, Fort Lee Gazette, *assigned to cover* the U.S. politics. During one of his vacations in India, Paul met rarely seen Himalayan masters. Then, *while on vacation,* he wrote a very interesting article on the "Saints of the East" being inspired by the yogis of India. Clearly, Paul created this work "outside" the scope of his employment with the Fort Lee Gazette. Consequently, this creative work is NOT a *work made for hire* and Paul should be named as *the author of the work* in the application for *copyright* registration.

   Example-2. Jill is an expert staff-photographer, working for the *International Geographic* monthly magazine, *designated to cover* under-the-water sea world. Once while Jill was on a *personal trip* to Hawaii to meet up with family, she happened to visit the Hawaii Volcanoes National Park. She captured the natural eruptions of Mauna Loa and Kilauea in all their splendor using her *personal* photographic equipment and wrote a book on it including her photographs. Clearly, this creative work of Jill falls "outside" the scope her employment with the *International Geographic*. Therefore, Jill's work is NOT a *work made for hire,* and Jill should be named as *the author of the work.*

3. *Work that is specially ordered or commissioned*

   Example-1. Roger is a Professor of Chemistry at the University of America (UA). The UA commissioned Roger: (a) *to set the chemistry test* and *prepare the answer material for the test* for their school's National Entrance Exam; further, (b) UA and Roger *entered into an express written agreement,* which (c) *stated that the above transaction is a work made for hire;* finally, (d) *both the parties signed the agreement* before the work started. This case perfectly exemplifies *the work that is specially commissioned,* as it satisfies all the four requirements: *first,* the test and answer material fall within the nine categories of works made for hire; *second,* UA and Roger enter into an express written agreement; *third,* the agreement clearly states that it is a *work made for hire;* and, *fourth,* the agreement was signed by both parties before the work started.

(*Continued*)

CLARIFICATION AND EXAMPLES (*Continued*)

Example-2. Debbie and Rose worked as employees at the Artificial Flowers company, before Rose retired from the firm. Debbie knew that Rose was an expert in *creative* floral bouquets (3-D art design). On behalf of the Artificial Flowers, Debbie ordered Rose to make 50 *creative* floral bouquets for a flat fee, under a signed agreement. After the work was complete, Debbie decided to apply for a *copyright* registration of this work. She was surprised to learn that in the application for a *copyright* registration, Rose (the independent contractor) should be named as *the author of the work,* not the Artificial Flowers, and that the *work made for hire* box should be checked as "No." Thus, in this case, *though there was an express signed written agreement,* the work itself does not qualify as a *work made for hire* because the 3-D *art design does not fall within any of the nine categories.*

4. **Derivative works.** The U.S. Copyright Law defines a "derivative work" as:

> A work based upon one or more pre-existing works, such as a translation, musical arrangement, dramatization, fictionalization, motion picture version, *sound recording,* art reproduction, abridgment, condensation, or any other form in which a work may be recast, transformed, or adapted. A work consisting of editorial revisions, annotations, elaborations, or other modifications, which, as a whole, represent an original work of authorship, is a *derivative work.* 17 U.S.C. § 101.[46a,b]

A *derivative work* is created by the process of recasting, transforming, or adapting one or more pre-existing works.

The "derivative works" contain two distinct forms of authorship:

- The *authorship* in the pre-existing work(s) that has been recast, transformed, or adapted within the *derivative work,* and
- The *new authorship* involved in recasting, transforming, or adapting the pre-existing work(s).

It must be noted that *copyright infringement* results: when a *copyrighted* work is …… *made into a derivative work without the permission of the copyright owner.*[46c] A *derivative work* can be either a "new version" (movie) of a pre-existing work (novel) or a "derived work" from it (translation).

The author of a *derivative work* (that recasts, transforms, or adapts a pre-existing work), may claim a *copyright in the derivative work,* provided it uses the pre-existing material in a lawful manner. A *derivative work's copyright* covers *only* the additional material that the author brought into the work and *not* any of the pre-existing material that may still be a part of the *derivative work.*[46d] Hence, a *copyright* in the "new version" *has absolutely no effect* on the *copyright* or public domain status of the pre-existing material.[46b]

The *new authorship* of the *derivative work* can also be registered, provided that it contains a sufficient amount of original authorship. A *derivative work's registration* does not cover any pre-existing material, previously registered material, public domain material, or third-party material that appears in the work.

Stated differently:

> The *copyright in a derivative work* is independent of, and does not affect or enlarge the scope, duration, ownership, or subsistence of, any *copyright* protection in the pre-existing material. 17 U.S.C. § 103(b).[46e]

5. **Compilations**. The U.S. Copyright Law defines a "compilation" as:

> A work formed by the collection and assembling of pre-existing materials or of data that are selected, coordinated, or *arranged in such a way that the resulting work as a whole constitutes an original work of authorship." The term compilation includes collective works*. 17 U.S.C. § 101.

A *compilation* results from a process of selecting, bringing together, organizing, and arranging previously existing material of all kinds, regardless of whether the individual items in the material have been or ever could have been subject to *copyright.*[47a]

Examples of *compilations* include: a directory of fire & emergency services in the Eastern United States; a list of the best seller books of 2017; and a collection of the best paintings during the World War II years.

The author of a compilation may claim a *copyright* in an original selection, coordination, and/or arrangement of pre-existing material, provided that the material has been used in a lawful manner. A *copyright in a compilation* "extends only to the material contributed by the author of such work" and does not "imply any exclusive right in the pre-existing material." 17 U.S.C. § 103(b).[46e] Further, a *copyright in a compilation* does NOT protect the data, facts, and/other *uncopyrightable* materials in the work.[47b] Similarly, *a registration for a compilation* also does NOT cover any of the previously published material, previously registered material, public domain material, or third-party material that appears in the compilation. In fact, the *copyright* in a factual compilation is really very thin. In fact, a compiler may be free to use the facts contained in a previous compilation, without copying the same selection and arrangement. For the USCO's practices and procedures on the *copyrightability of compilations,* and guidance in preparing an *application to register a compilation,* see elsewhere.[47c,d]

6. **Collective works**. The U.S. Copyright Law defines "collective work" as:

 A work, such as a periodical issue, anthology, or encyclopedia, in which a number of contributions, constituting separate and independent works in themselves, are assembled into a collective whole, 17 U.S.C. § 101.[48a]

 The term "compilation" includes collective works. Thus, *collective works* must comply with the statutory requirements for *compilations.* In order to establish itself as a *collective work,* it must involve sufficiently creative selection, coordination, or arrangement of the *component works.*

 A *collective work* must contain *multiple contributions (component works).* Works, such as shown below, containing "relatively few separate elements" do not meet this requirement[48b]:

 - A work containing a single contribution;
 - A composition that merely consists of words and music;
 - A publication that merely combines;
   - A single work with illustrations or front matter; and
   - Two one-act plays;

 The rule of thumb here is that a *component work* that is incorporated in a *collective work* must itself qualify as a *separate and independent work."*[48c] Stated differently, any contribution to a *collective work* must be *an original work of authorship* that is eligible for *copyright protection* under Section 102(a) of the Copyright Act, regardless of whether that contribution is currently protected or whether the *copyright* in that contribution has expired.[48d]

 As stated above, preparing a *collective work* involves: assembling or gathering of *separate and independent works... into a collective whole.*[48e] Consequently, *collective works* contain two distinct forms of authorship:

 - The *compilation authorship* in the creation of the collective work, involving—selection, coordination, and/or arrangement of a number of separate and independent works and assemblage into a collective whole; and

- The *authorship in the separate and independent works* included in the *collective work*, such as a research article in a journal or a poem in an anthology.

An applicant may *register* the *collective work* containing *multiple, separate*, and *independent works* provided: (i) the *copyright* in the *collective work* and the *component works* are owned by the *same claimant*, and (ii) the *component works* have not been previously published, registered, or in the public domain.

7. **Anonymous works**. According to the U.S. Copyright Law, a work is an "anonymous work" if "no natural person is identified as the author on the copies or phonorecords of the work," 17 U.S.C. § 101.[49a]

   Clearly, this statute implies that anonymous works *are limited only to works (copies or phonorecords) created by natural persons (human beings).*

   It is important to note that specifying the author's name while registering an *anonymous work* creates *a clear record of authorship and ownership of the copyright*, and it may extend or reduce the term of the *copyright*, depending on the circumstances. A few pointers may be helpful in this regard. First, the applicant should check the *anonymous box* only if the author is a human being; second, if the author is a corporation, limited liability company, partnership, or other legal entity, the author's full name should be provided in the Name of Author field/space; third, when the author's name appears on the copies or phonorecords, the work will not be considered as an *anonymous work*, even if the author does not wish to reveal his or her identity in the registration record; fourth, if the applicant checks the *anonymous box* or asserts that the *author wishes to remain anonymous*, the application may be questioned if the author turns out to be a legal entity; and finally, if the author's name does not appear on the copies or phonorecords of the work, the applicant is not required to provide the author's name in the application. Instead, the applicant may leave the Name of the Author field/space blank and check the box marked "Anonymous."

8. **Pseudonymous works**. According to the U.S. Copyright Law, a work is considered as a "pseudonymous work" if *the author is identified under a fictitious name on the copies or phonorecords of the work*, 17 U.S.C. § 101.[49]

   This statute implies that *pseudonymous works are limited only to works created by an individual*. It must be noted that the USCO will accept only a name for a *pseudonym* and NOT a number or symbol.

   It is important to know that, while registering a *pseudonymous work*, providing the author's real name creates a clear record of *authorship and ownership of the copyright*, and it may extend or reduce the term of the *copyright*, depending on the circumstances. Hence, while registering a *pseudonymous work*, an applicant is encouraged to provide the author's real name in the application, even if the author's name does not appear on the copies or phonorecords of the work. Alternately, an applicant may provide the *author's full name* and the author's *pseudonym*, provided that the application clearly indicates which is the real name and which is the pseudonym (e.g., "Samuel Clemens," the real name, and "Mark Twain," the pseudonym).

Duration of Copyright Protection

Indeed, the U.S. Copyright Law dealing with the *duration of copyright protection* is complex. In fact, varying standards apply based on whether the federal statutory *copyright* protection was received *before* or *on/after January 1, 1978*, the date when the current law, the Copyright Act of 1976, came into effect. Further, the duration of a *copyright* is also impacted by the many amendments enacted since January 1, 1978.[50]

1. **Works securing first federal statutory protection in United States on/after January 1, 1978**

   For *all the works that fall into this category*, the Copyright Act of 1976, as amended in 1998, establishes *a single copyright term*, but *different methods* for computing the *duration of a copyright*. These works are of two kinds:

   Works created on or after January 1, 1978

   The law automatically protects *a work that is created and fixed on or after January 1, 1978*, from the moment of its creation.

   The duration of a *copyright* varies for different works as shown below:

   a. *Individual works*. The term is, author's life plus an additional 70 years;
   b. *Joint works by two or more authors who did not work for hire*. The term lasts for 70 years after the last surviving author's death;
   c. *Works made for hire, anonymous*, and *pseudonymous works*. The duration of a *copyright* is 95 years from first publication or 120 years from creation, whichever is shorter (unless the author's identity is later revealed in Copyright Office records, in which case the term becomes the author's life plus 70 years).

Works in existence, but not published or copyrighted on January 1, 1978

The law also gives automatic federal *copyright* protection *to works that were created before January 1, 1978, but not published or registered*. The duration of a *copyright* in these works is also computed the same way as described above: life + 70 years, 95, or 120 years, depending on the nature of authorship. First of all, all works in this category *are guaranteed a minimum of 25 years of statutory protection*. The law further assures that under no circumstances, would a *copyright* in a work in this category have expired before December 31, 2002. Furthermore, if a work in this category got published before December 31, 2002, *the term would extend by another 45 years*, through the end of 2047.

2. **Works already under federal statutory protection in the United States before 1978**

   For works that already secured statutory *copyright* protection in the United States before January 1, 1978, the 1976 Copyright Act retains the system in the previous *copyright* law—the Copyright Act of 1909—for *computing the duration of protection*, but with certain changes. These and other details can be found elsewhere.[50]

Plagiarism and Copyright Infringement

It is very important to understand the concepts of *plagiarism* and *copyright infringement*.

In earlier discussion, we have clarified the scope of the six exclusive rights of the *copyright owner* and *copyright infringement* with suitable examples. There is a useful resource if one is interested in understanding this subject matter in a thorough manner. See 17 U.S.C. §§ 501–513 for details.

In this section, however, the goal is to explain the differences between *plagiarism* and *copyright infringement*, which are sometimes misunderstood as *similar* or *same* ideas. One only needs to do a quick literature search to find excellent references that can clarify the differences between the two concepts and dispel this misconception.[51a–e]

Plagiarism and *copyright infringement* are two different kinds of violations. One can *plagiarize* without committing *copyright infringement*; similarly, one can commit *copyright infringement* without *plagiarizing*. Furthermore, *plagiarism* and *copyright infringement* can occur, sometimes simultaneously. Figure 4.1 illustrates these ideas very well.

There is no universal definition for *plagiarism*. However, there are many similar definitions, as shown below: "*Plagiarism* is using someone else's work

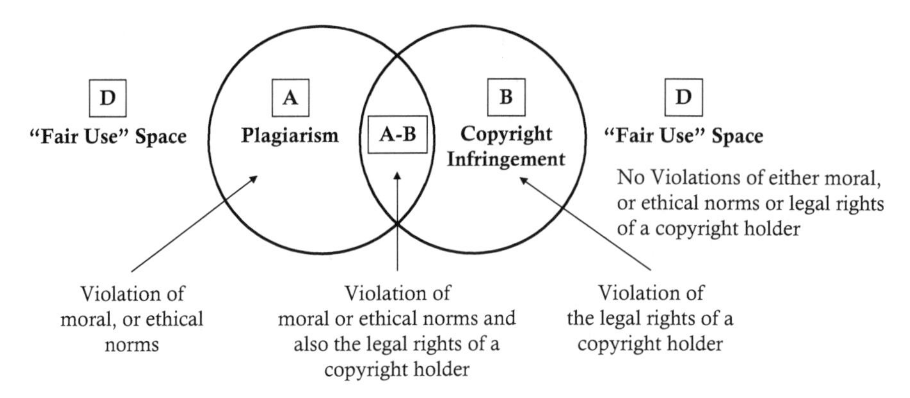

FIGURE 4.1
The spaces of plagiarism, copyright infringement, and fair use.

or ideas, without giving proper credit";[51a] "*Plagiarism* occurs when you use someone else's words or ideas without giving credit to the original author,";[51b] "*Plagiarism* is using someone else's work without giving proper credit—a failure to cite adequately";[51c] "*Plagiarism* is defined as taking the original work or works of another and presenting it as your own"[51d]; Finally, "*Plagiarism* is, simply, taking someone else's work and passing it off as your own."

From these definitions, three elements are critical to *plagiarism*: (a) taking/using someone else's work/ideas; (b) not giving credit; and (c) passing the ideas as one's own. Simply put, the *three violations* committed here are—*using* (some say, *stealing*) *someone else's intellectual property, failing to cite* (some say, *credit*) *the source(s),* and *reporting* (some say, claiming) *as though the work/idea(s) belongs to oneself.* Therefore, *plagiarism* involves *violations of moral or ethical norms.*

In contrast, *copyright infringement* involves *violations of the legal rights of a copyright holder.* According to 17 U.S.C. § 501(a):

> *Anyone who violates any of the exclusive rights* of the *copyright* owner as provided by sections 17 U.S.C. § 106 through 17 U.S.C. § 122 or of the author as provided in section 106A(a), or who imports copies or phonorecords into the United States in violation of section 602, *is an infringer of the copyright or right of the author,* as the case may be.[26a]

Simply stated: *Copyright infringement occurs* when a *copyrighted* work is reproduced, distributed, performed, publicly displayed, or made into a *derivative work,* without the permission of the *copyright* owner.[26b] In this regard, there are four possibilities (see, spaces A, B, AB, and D)[51b] (Figure 4.1):

1. **Plagiarism without *copyright* infringement (space A).** Example: Jack is a high school student who blatantly copied *key ideas* from the book of "A Tale of Two Cities," written by Charles Dickens, in his report on *Societal Impact of Revolutions,* and submitted his work to

the teacher, *concealing the truth that the ideas are not original* and *without acknowledging the work of* Charles Dickens. Is this *plagiarism* or *infringement*? Indeed, this is *plagiarism* because Jack copied Charles Dicken's *work/ideas*, and *presented them as his own, without citing/acknowledging the original work*. However, this is not *copyright infringement* because the *copyright* in the "A Tale of Two Cities" has expired, and the work is in the public domain. Moreover, a *copyright does not protect* the ideas, 17 U.S.C. § 102(b). Thus, this is an example of *moral and ethical violations*.

2. ***Copyright* infringement without plagiarism (space B).** Example: Stewart is a business school professor. He was approached by some local industries in his town to create a 3-Day training course on *business leadership* for entry-level company executives. Accordingly, Stewart prepared three books on *Business Leadership* for this course, each containing photocopies of ten book chapters taken from different books, *the authors of which he carefully cited at the end of each book*. However, *he did not seek any permission from the authors for this work*. In addition, Stewart collected $1000 admission fee from each executive for this training course, and gave the books free, as part of the course package. Is this *plagiarism* or *infringement*? Indeed, this is a clear case of *copyright infringement* and not *plagiarism*, because: (a) first, Stewart engaged in *unauthorized photocopying* of book chapters (violation of reproduction right); (b) second, he engaged in an *unauthorized compilation* (violation of adaptation right) and *distribution* of it (violation of public distribution right); and (c) finally, he *gave credit to all the authors and their contributions at the end of each book*. Thus, this is an example for *infringement* (i.e., *legal violations of the exclusive rights of a copyright holder*). See, 17 U.S.C. § 501(a).
3. **Simultaneous violations of plagiarism and *copyright* infringement (space AB).** Example: Joseph is a university professor in organic chemistry. As he is a renowned expert in alkaloids research, he was invited to deliver a plenary lecture at the North America Convention for Organic Chemists. During his 1-hour lecture, he showed fascinating new ideas for future research, based on ten *professionally prepared figures* that caught everyone's attention. As it turned out, the ten *professionally prepared figures*, that Joseph showed in his talk, were *copied verbatim out of the works of other lesser known researchers*. In fact, Joseph *neither sought the permission of those authors to use their works in his talk, nor made any effort to acknowledge those authors and point out the ideas shown in the figures came from those authors*. Is this *plagiarism* or *infringement*? Unfortunately, this is a case of *both*. Firstly, it is *copyright infringement* because Joseph "copied multiple figures out of the works of other authors without permission" (violation of reproduction right) and "displayed them

without authorization" (violation of public display right). In addition, it is *plagiarism* because Joseph presented the talk "without pointing out who the real authors of those slides were," and "giving the false impression that the ideas in those figures were his own," (violation of moral and ethical norms).

4. **Fair use space (space D).** Example: Let us extend the same example discussed above. Joseph, the organic chemistry professor, taught a course on "The Chemistry of Alkaloids" to graduate students at his university. Let us suppose that he *included the* same ten *professionally prepared figures* in his class lectures. *but did two things differently.* First, he *spread them over 20 classroom lectures* throughout the semester; second, he *cited the names of all the authors* and *let the students know that these figures were taken from the literature.* He was making much bigger points, while he was using these figures. Does this constitute *plagiarism* or *infringement*? Indeed, the answer in this case is, "There is neither *plagiarism* nor *copyright infringement*." We can regard this as, "fair use," as it falls under *fair use doctrine*.[52]

Some final observations: *Plagiarism* is a violation against *the author of an original work,* while *copyright infringement* is a violation against *the copyright owner.* In the USA, *plagiarism* is unethical and/immoral, but NOT illegal. On the other hand, *copyright infringement* is definitely illegal, but can ALSO be unethical and/immoral if *copyright* violations are proven to be perpetrated *willingly* and *with full knowledge.* That is why *plagiarism* battles are fought *in the courts of public opinion,* whereas the wars of *copyright infringement* are fought *in legal courts.* One can avoid *plagiarism* by "properly acknowledging the authors of original works from where the ideas/concepts are borrowed and giving due credit to their intellectual contributions." On the contrary, *copyright infringement* can be avoided only by "understanding the *copyright* laws and complying with them, knowing the legal consequences of *copyright* violations vs. fair use."

References

1. https://www.eucopyright.com/en
2. https://www.copyright.gov/circs/circ01.pdf
3. (a) https://www.copyright.gov/title17/title17.pdf; and (b) https://www.copyright.gov/title17/92chap1.html
4. (a) https://www.copyright.gov/comp3/docs/compendium-12-22-14.pdf; and (b) https://www.copyright.gov/title17/92chap1.html#102

5. (a) https://www.copyright.gov/title17/92chap1.html#102; (b) https://www.copyright.gov/circs/circ33.pdf
6. https://www.copyright.gov/title17/92chap1.html#101
7. https://copyright.gov/comp3/chap800/ch800-performing-arts.pdf
8. https://www.copyright.gov/title17/92chap1.html#117
9. (a) https://www.copyright.gov/circs/circ50.pdf; and (b) https://www.copyright.gov/circs/circ56a.pdf
10. (a) https://www.copyright.gov/circs/circ50.pdf (b); and (b) https://www.copyright.gov/title17/92chap1.html#114
11. (a) https://www.copyright.gov/title17/92chap1.html#101; and (b) https://www.copyright.gov/circs/circ45.pdf
12. (a) https://copyright.gov/comp3/chap800/ch800-performing-arts.pdf; and (b) https://www.copyright.gov/title17/92chap1.html#113
13. https://copyright.gov/comp3/docs/glossary.pdf
14. (a) https://www.copyright.gov/title17/92chap1.html#103; and (b) The language used in the Reference 14 (a) above is essentially retained to avoid misrepresentation.
15. (a) https://www.copyright.gov/title17/92chap1.html#104; and (b) The language used in the Reference 15 (a) above is essentially retained to avoid misrepresentation; (c) https://www.copyright.gov/title17/92chap1.html#105; (d) https://copyright.gov/history/comp/compendium-two.pdf
16. https://law.justia.com/cases/federal/appellate-courts/F2/945/509/289784/
17. https://copyright.gov/history/comp/compendium-two.pdf
18. https://copyright.gov/comp3/chap800/ch800-performing-arts.pdf
19. (a) https://www.copyright.gov/title17/92chap1.html#102(b); (b) https://www.copyright.gov/circs/circ33.pdf; and (c) https://www.copyright.gov/title37/202/37cfr202-1.html
20. https://www.copyright.gov/circs/circ33.pdf
21. https://www.copyright.gov/title17/92chap1.html#106
22. (a) https://www.copyright.gov/title17/92chap1.html#107; and (b) https://www.copyright.gov/title17/92chap1.html#108-122
23. WIPO Lex: http://www.wipo.int/wipolex/en/
24. https://www.copyright.gov/title17/92chap1.html#107
25. https://ogc.harvard.edu/pages/copyright-and-fair-use%20
26. (a) https://www.copyright.gov/title17/92chap5.html; and (b) https://www.copyright.gov/help/faq/faq-definitions.html
27. https://www.avvo.com/legal-guides/ugc/copyright-infringement-litigation-faq
28. For definitions of copies/phonorecords, see 4.3.1.
29. McJohn, S. M. (2009). *Intellectual Property; Examples & Explanations*. New York: Aspen Publishers, p. 145.
30. For a definition of derivative works, see 4.2.2.1.
31. https://www.bitlaw.com/copyright/scope.html
32. https://www.copyright.gov/title17/92chap1.html#109
33. http://ir.lawnet.fordham.edu/cgi/viewcontent.cgi?article=1094&context=iplj
34. https://copyright.gov/comp3/chap800/ch800-performing-arts.pdf
35. (a) http://invention-protection.com/ip/publications/docs/Public_Performance_of_Copyrighted_Material.html; and (b) https://www.copyright.gov/title17/92chap5.html

36. (a) http://goodattorneysatlaw.com/right-to-public-display/; (b) http://www.law.uci.edu/faculty/full-time/reese/reese_illinois.pdf; and (c) RAM Copy Doctrine suggests that "accessing a work on a computer *infringes the reproduction right because it utilizes temporary storage in the computer's RAM.*"
37. (a) https://www.gpo.gov/fdsys/pkg/PLAW-104publ39/pdf/PLAW-104publ39.pdf; (b) https://en.wikipedia.org/wiki/Digital_Performance_Right_in_Sound_Recordings_Act; (c) Martin, R. (1996). The digital performance right in the sound recordings act of 1995: Can it protect U.S. sound recording copyright owners in a global market? *Cardozo Arts and Entertainment Law Journal.* 14: 733; (d) Cohen, J.; Loren, L.; Okediji, R.; and O'Rourke, M. (2006). *Copyright in a Global Information Economy.* New York: Aspen Publishers, pp. 466–467; (e) Sen, S. (2007). The denial of a general performance right in sound recordings: A policy that facilitates our democratic civil society? *Harvard Journal of Law and Technology.* 21(1): 262; and (f) https://www.copyright.gov/legislation/dmca.pdf
38. https://www.eucopyright.com/en/why-should-i-register-my-work-if-copyright-protection-is-automatic
39. http://www.wipo.int/copyright/en/activities/copyright_registration/
40. http://www.wipo.int/wipolex/en/treaties/text.jsp?file_id=278718
41. (a) https://www.copyright.gov/comp3/chap500/ch500-identifying-works.pdf; (b) https://www.copyright.gov/fls/sl35.pdf; (c) https://www.copyright.gov/comp3/chap200/ch200-registration-process.pdf; and (d) See the following cases: "Although registration is 'permissive,' both the certificate and the original work must be on file with the Copyright Office before a copyright owner can sue for infringement." *Petrella v. Metro-Goldwyn-Mayer, Inc.*, 132 S. Ct. 1962, 1977 (2014). "Though an owner has property rights without registration, he needs to register the copyright to sue for infringement." *Alaska Stock, LLC. v. Houghton Mifflin Harcourt Publishing Co.*, 747 F.3d 673, 678 (9th Cir. 2014).
42. https://www.copyright.gov/comp3/chap2200/ch2200-notice.pdf
43. (a) https://www.copyright.gov/comp3/chap500/ch500-identifying-works.pdf; and (b) https://copyright.gov/comp3/chap600/ch600-examination-practices.pdf
44. (a) See Reference 7; (b) See Reference 43 (a), pages 12–13; (c) H.R. Rep. No. 94–1476, at 120, reprinted in 1976 U.S.C.C.A.N. at 5736; (d) S. Rep. No. 94-473, at 103–04; and (e) https://www.copyright.gov/title17/92chap2.html
45. https://www.copyright.gov/comp3/chap500/ch500-identifying-works.pdf
46. (a) https://www.copyright.gov/title17/92chap1.html; (b) See Reference 45; see also, https://www.copyright.gov/circs/circ14.pdf; (c) See Reference 26(b); (d) See H.R. 94 1476, at 57, reprinted in 1976 U.S.C.C.A.N. at 5670; S. Rep. No. 94-473, at 55; and (e) https://www.copyright.gov/title17/92chap1.html#103
47. (a) H.R. Rep. No. 94-1476, at 57, reprinted in 1976 U.S.C.C.A.N. at 5670; S. Rep. No. 94-473, at 55; (b) See, the case of *Feist Publications, Inc. v. Rural Telephone Service Co.*, 499 U.S. 340, 360 (1991) (which states, "the copyright in a compilation does not extend to the facts it contains"); (c) https://copyright.gov/comp3/chap300/ch300-copyrightable-authorship.pdf; and (d) https://copyright.gov/comp3/chap600/ch600-examination-practices.pdf

48. (a) https://copyright.gov/comp3/chap800/ch800-performing-arts.pdf; (b) H.R. Rep. No. 94-1476, at 122, reprinted in 1976 U.S.C.C.A.N. at 5737; S. Rep. No. 94-473, at 105; (c) H.R. Rep. No. 94-1476, at 122, reprinted in 1976 U.S.C.C.A.N. at 5737; S. Rep. No. 94-473, at 105; (d) https://www.copyright.gov/title17/92chap1.html#102; and (e) H.R. Rep. No. 94-1476, at 120, reprinted in 1976 U.S.C.C.A.N. 5659, 5736; S. Rep. No. 94-473, at 104 (omission in original)
49. https://copyright.gov/comp3/chap600/ch600-examination-practices.pdf
50. https://www.copyright.gov/circs/circ15a.pdf
51. (a) https://researchguides.uic.edu/c.php?g=252209&p=1682805; (b) https://library.osu.edu/document-registry/docs/587/stream; (c) http://www.plagiarismchecker.com/plagiarism-vs-copyright.php; (d) https://www.plagiarismtoday.com/2013/10/07/difference-copyright-infringement-plagiarism/; and (e) https://sarafhawkins.com/difference-copyright-and-plagiarism/
52. For "Fair Use Doctrine," see, Section 4.5.1.

5

Trademarks

An *Identification mark* is any *sign* or *other indication* that serves to *identify* someone or something.

Collins English Dictionary

Identification marks are important for businesses for several reasons: first, they help to provide *instant recognition* and *unique identity* to every business and its products or services; second, they help the businesses to create and build demand for their products and services; third, they enable customers to quickly identify and make a purchase based on a recognized mark; and, finally, they enable businesses to earn the confidence, trust, and loyalty of the customers in the long run.

In America, the United States Patent and Trademark Office (USPTO) registers two types of *identification marks*, namely—"trademarks" and "service marks" for *products* and *services* respectively, to support the entrepreneurs, companies, and customers in their business pursuits, and contribute to the economic growth and progress of the nation. The Commissioner for Trademarks is the head of the United States (U.S.) Trademark Organization who ensures that USPTO examines all trademark applications and grants registrations when applicants are entitled to them; records ownership changes of trademarks; maintains search files and records of U.S. trademarks; and publishes and disseminates trademark information. This chapter attempts to cover the essential concepts as well as business-useful information on *trademarks*, *service marks*, and *other types of marks*.

Trademarks and Service Marks—Basic Requirements

A "trademark" is *a word, phrase, symbol, or design, or a combination thereof, that identifies and distinguishes the source of the goods of one party from those of others.*[1a,b] A *service mark* is *the same as a trademark, except that it identifies and distinguishes the source of a service rather than goods.*[1a,b]

According to the U.S. Trademark Law[2a] (15 U.S.C. § 1127):[2b]

> The term *trademark* includes *any word, name, symbol, or device, or any combination* thereof –
>
> a. Used by a *person;* or
> b. Which a *person* has *a bona fide intention to use in commerce* and applies to register on the principal register established by this chapter, *to identify and distinguish his or her goods,* including a unique product, *from those manufactured or sold by others and to indicate the source of the goods,* even if that source is unknown.

According to the U.S. Trademark Law[2a] (15 U.S.C. § 1127):[2b]

> The term *service mark* means *any word, name, symbol, or device, or any combination* thereof
>
> a. Used by a *person;* or
> b. Which a *person* has *a bona fide intention to use in commerce* and applies to register on the principal register established by this chapter, *to identify and distinguish the services of one person,* including a unique service, *from the services of others* and *to indicate the source of the services,* even if that source is unknown.

Thus, to start with, we must know that "trademark" and "service mark" are similar, except that the former is *used to identify and distinguish the source of goods* of one party from those of the others, while the latter is used *to do exactly the same for services.*

Next, we must know the definitions of various terms in U.S. Trademark Law. Thus, the U.S. Trademark Law, defines a "person" very broadly. The definition of a "person" (see 15 U.S.C. § 1127)[2b] includes—a *natural person,* as well as a *juristic person.* A *juristic person* includes a firm, corporation, union, association, or other organization capable of suing and being sued in a court of law. In addition, a "person" includes—the United States, any U.S. state, any agency or instrumentality thereof, or anyone representing the United States and its interests.

Next, "commerce" means all commercial activities lawfully regulated by the U.S. Congress. In the U.S. Trademark Law, the term "use in commerce" means *the bona fide use of a mark in the ordinary course of trade, and not made merely to reserve a right in a mark.*[2b]

Finally, we need to know what the U.S. Trademark means by *any word, name, symbol, or device, or any combination.* Thus, trademarks and service marks use the following *words, names, symbols, devices, and combinations* in order to *identify and distinguish the source of the goods/services of one party from those of others:*[3,4a,b]

Letters and Abbreviations

The letter **M** for McDonalds *fast foods (products)*; **f** for Facebook *services*; the abbreviations of **GE** for General Electric *products*; **IBM** for the *products & services* of International Business Machines; **BBC** for British Broadcasting Corporation *services*; **FedEx**® for Federal Express Corporation *services*; **UPS** for United Parcel Service *logistics services*; **VW** for Volkswagen *automobiles (products)*; **BMW** for the *ultimate driving machines (products)* from Bavarian Motor Works; and **Infosys**® for *information systems (services).*

Words, Names, Numbers, and Logos

Coca-Cola for *carbonated soft drinks*; **Gatorade** for *sports drinks*; ***Dove*** for *personal care products*; ***Nike*** *for shoes*; **LEVI'S 501** for blue jeans; **Google** for Internet *products & services*; ***VISA*** for *credit cards*; **Honda** for *automobiles*; **SEIKO** for *watches*; **Samsung** for *electronics*; **Boeing 787** for *airplanes*; the logo of a *multicolored peacock* for **NBC Television**; the logo of a *munched* apple for **Apple** products; the logo of *a birdie* for **Twitter**; the logo of *a sea-shell* for **Shell Oil & Gas Company** products; and the logo of a ringing bell for **Taco Bell**.

Pictures and Drawings

Polo apparel, accessories, and fragrances—the *picture of a polo player*; **Burberry** luxury fashion house—the *picture of a knight on a horse*; ***Nike*** shoes and other products—the *tick mark drawing* (√); and **100% Columbian Coffee**—the *picture of a triangular logo with a man and horse.*

Letters, Logos, Abbreviations, and Words with Slogans

General Mills—the letter **G** positioned over the words, General Mills; **McDonalds**—the letter **M** positioned over the slogan, *I'm lovin' it*; ***Nike***—the logo of a tick mark (√) positioned over the slogan, *Just Do It*; HSBC—the abbreviation HSBC positioned over the slogan, *The world's local bank*; **Accenture**—the word positioned over the slogan, *High performance. Delivered*; and **Alibaba.com**—the words positioned over *Global Trade Starts Here.*

Words, Shapes, and Colors

Walt Disney World in a unique font and *black color*; **Google** in a unique font *with multi-colors*; **Coca-Cola** in a unique font *in red color*; **Twitter** in a unique font *with a blue birdie*; **LEGO** in a unique font in *red, white,* and *yellow*; and ***SUBWAY*** in a unique font *in black, white,* and *yellow.*

Product Shape

Coca-Cola bottle for its unique shape; the **iPod** for its unique shape; and the **Perrier** sparkling water bottle also.

Sounds/Jingles[5]

Some well-known examples are: *Nationwide is on your side* by Nationwide Insurance; *I'm lovin' it* by McDonalds; and *Like a good neighbor, State Farm is there* by State Farm insurance.

Other Categories of Marks—Basic Requirements

There are also other kinds of *identification marks,* intended for different purposes.

Collective Marks

They are a particular type of *trademark* or *service mark,* intended to indicate the *membership in a union, an association, or other organization* or *distinguish the goods or services of its member organizations from the non-members.*

The legal definition of a "collective mark" (15 U.S.C. § 1127[2b]) is as follows:

> The term *collective mark* means a *trademark* or *service mark*:
> 1. Used by the members of a *cooperative, an association, or other collective group or organization;*
> 2. Which such *cooperative, association,* or *other collective group* or *organization* has *a bona fide intention to use in commerce* and applies; and
> 3. To register on the *principal register* established by this chapter and *includes marks indicating membership in a union, an association, or other organization.*

"Collective marks" are used for two different purposes:[6a,b]

- To *indicate membership* of a cooperative, an association, or other collective group or organization; these are referred to as *collective membership marks;* and
- To *distinguish the geographical origin, material, mode of manufacture, quality, or other common characteristics of goods or services* of different enterprises; these are referred to as *collective trademarks and service marks.*

Some examples of *collective membership marks* include—**teamsters**,[7a] the labor union in the United States and Canada; **AFL-CIO**,[7b] the American Federation of Labor and Congress of Industrial Organizations, the largest federation of unions in the United States; **ACLU**,[7c] the American Civil Liberties Union, a non-partisan non-profit organization.

According to the World Intellectual Property Organization (WIPO):[6b]

> *Collective marks* are often *used to promote products which are characteristic of a given region*. In such cases, the creation of a *collective mark* has not only helped to market such products domestically and occasionally internationally, but has also provided a framework for cooperation between local producers. The creation of the *collective mark*, in fact, must go hand in hand with the development of certain standards and criteria and a common strategy. In this sense, *collective marks* may become powerful tools for local development.
>
> Consider, in particular products which may have *certain characteristics which are specific to the producers in a given region, linked to the historical, cultural, and social conditions of the area*. A *collective mark* may be used *to embody such features and as the basis for the marketing of the said products*, thus benefiting all producers.

Some examples of *collective trademarks and service marks* include—**NRA**,[8a] the National Rifle Association of America that advocates for gun rights; **NCC**,[8b] the National Cotton Council of America, that ensures the ability of all U.S. cotton industry segments to compete effectively and profitably in the raw cotton, oilseed, and U.S.-manufactured product markets at home and abroad; **IUPAT**,[8c] the International Union of Painters and Allied Trades, a union representing the interests of construction specialists, including—painters, glaziers, wall coverers, flooring installers, convention and trade show decorators, glassworkers, sign and display workers, asbestos worker/hazmat technician, and drywall finishers in the United States and Canada; **NCA**,[8d] the National Cleaners Association, dedicated to the welfare of the well-groomed consumers and the professional cleaners and suppliers serving them; and **PCPC**,[8e] the Personal Care Products Council, the leading national trade association representing the global cosmetic and personal care products industry.

Certification Marks

"Certification marks" are *marks which a business can use, when it meets the standards set by the mark's owner*. Therefore, the mark's owner has the responsibility to ensure that the businesses *that use the mark meet the standards*, and *that the consumers understand what the mark stands for*.

In the United States of America (USA), the legal definition of a "certification mark" (15 U.S.C. § 1127[2b]) is as follows:

> *Certification mark* means *any word, name, symbol, or device, or combination* thereof:
>
> a. Used by *a person other than its owner*; or
> b. Which its owner has *a bona fide intention to permit a person other than the owner to use in commerce* and files an application to register on the principal register established by this chapter, *to certify regional or other origin, material, mode of manufacture, quality, accuracy, or other characteristics of such person's goods or services* or that the work or labor on the goods or services was performed by members of a union or other organization.

According to the European Union Intellectual Property Office, a "certification mark" *is a new kind of trade mark* at the EU level, although some national level systems in the EU have used it for a few years.[9]

Thus, the European Union Intellectual Property Office defines a "certification mark" as:

> A *mark* that is *capable of distinguishing goods or services which are certified by the proprietor of the mark* in respect of material, mode of manufacture of goods, or performance of services, quality, accuracy, or other characteristics, *with the exception of geographical origin, from goods and services which are not so certified.*

Thus, a *certification mark* in the EU is employed to *guarantee specific characteristics of certain goods and services.*[9] Hence, the *goods and services* carrying a *certification mark* are expected to comply with the standard(s) specified in the regulations of use and governed by the control and responsibility of the *certification mark* owner; this is true, *regardless of the identity of the undertaking that actually produces or provides the goods and services* and actually *uses* the *certification mark.*

In the EU, a *certification mark* has the following *limitations*:

- It "cannot be owned" by a business person involved in the supply of goods and services *of the kind being certified;*
- It "cannot be used" by the owner of the mark for the *certified goods or services;* and
- It "cannot be filed" by anyone *for distinguishing goods or services certified in respect of the geographical origin.*

Some examples of *certification marks* include—the symbol of the **Underwriters' Laboratories, Inc.**,[10a,b] found on the labels and tags of certain appliances, tools, machinery, and material products indicates a certification by the

Underwriters' Laboratories, Inc. that these goods conform to the safety standards established by the Underwriters' Laboratories, Inc.; **ISI** mark,[10c] developed by the Bureau of Indian Standards, is a *certification mark* for industrial products in India; **ISO 9000** series of standards developed by the International Organization for Standardization, that define, establish, and maintain *an effective quality assurance system for manufacturing/service industries.*[10d]

Trade Dress

The U.S. Trademark Law defines a "symbol" or "device,"—*in the broadest of terms, almost anything that is capable of carrying the meaning, that human beings might use*—which qualify, as a *trademark* or *service mark*. This comprises a whole universe of—letters, words, numbers, shapes, 2-dimensional designs, 3-dimensional objects, colors, textures, scents, sounds, and any combination thereof.[11a]

"Trade dress" is a special type of *trademark* or *service mark* used for products or services or a combination of both. For products, "trade dress" is defined as, "its *total image* or *overall appearance* that may include features such as size, shape, color, or color combinations, texture, graphics, or even certain sales techniques." *John H. Harland Co. v. Clarke Checks, Inc.*, 771 F.2d 966, 980 (11th Cir. 1983), cited with approval in *Two Pesos, Inc. v. Taco Cabana, Inc.*, 505 U.S. 763, 112 S. Ct. 2753 (1992).[11b] For services, "trade dress" is defined even more broadly. In *Two Pesos, Inc. v. Taco Cabana, Inc.*, 112 S. Ct. 2753 (1992), the U.S. Supreme Court said *not only that restaurant decor may be protected as* "trade dress", *but also that restaurant—and other trade dresses may be inherently distinctive and protectable from the moment of adoption.*[11c]

Thus, the U.S. Supreme Court held protectable the "trade dress" (the restaurant atmosphere) of Taco Cabana, saying:[11c]

> Not only that a *festive eating atmosphere* having *interior dining and patio areas* decorated with artifacts, bright colors, paintings, and murals, the patio *includes interior and exterior areas* with the interior patio capable of being sealed off from the outside patio by overhead garage doors. The *stepped exterior* of the building is a festive and vivid color scheme using top border paint and neon stripes. Bright *awnings* and *umbrellas* continue the theme.

According to the Trademark Manual of Examining Procedures (TMEP):[11d]

> Trade dress constitutes a *symbol* or *device* within the meaning of §2 of the Trademark Act, 15 U.S.C. §1052. *Wal-Mart Stores, Inc. v. Samara Bros.*, 529 U.S. 205, 209-210, 54 USPQ2d 1065, 1065-66 (2000). Trade dress originally included only the *packaging* or *dressing* of a product, but in recent years has been *expanded to encompass the design of a product*. It is usually defined as the *total image and overall appearance* of a product, or the totality of the elements, and may include features such as size, shape, color or color combinations, texture, graphics. *Two Pesos, Inc. v. Taco Cabana, Inc.*, 505 U.S. 763, 764 n.1, 23 USPQ2d 1081, 1082 n.1 (1992).

An example of "trade dress" for a product is the *Coca-Cola bottle*, whose *shape* is distinctive and provides an instant consumer recognition to the beverage, though the *shape* itself serves no functional purpose (does not improve taste of the cola, preserve it better, or dispense the beverage easier, etc.). Another example is the *Ferrari car*, whose "trade dress" protects the brand from kit-car manufacturers offering replicas. (*Ferrari*) *Esercizio v. Roberts*, 944 F.2d 1235 (6th Cir. 1991).[11e] *Sounds* can be registered and protected under "trade dress," such as *the three-tone chime used by NBC*. Similarly, *smells* can be covered under "trade dress," such as the case of a manufacturer of thread and yarn for sewing, who registered it and protected its smell by the description, "high impact, fresh, floral fragrance reminiscent of plumeria blossoms." *In re Clarke*, 17 U.S.P.Q.2d 1238 (T.T.A.B. 1990).[11e] Because, "trade dress" is so broadly defined, it may run into the risk of *infringement*. However, by requiring the "trade dress" claimant to *precisely detail all the elements that constitute the* "trade dress," *protection from infringement may be possible.*

Trademarks and Service Marks—Substantive Requirements

In the preceding discussion, we have elaborated what *trademarks, service marks*, and *other categories of marks* are, as well as the *basic requirements* for their protection. In continuation, we shall outline below, the *substantive requirements* for their protection.

Thus, the *substantive requirements* for the protection of *trademarks, service marks*, and *other categories of marks*, are that:

- The marks *must* be "distinctive"; and
- The marks *must not fall into the categories excluded by the* U.S. Trademark Law, such as — "be confusing, deceptive, functional, scandalous, or disparaging of a group."

The "distinctiveness" of a *trademark* or *service mark* (a letter, word, number, phrase, slogan, logo, color, smell, sound, shape, package design, or a combination of all of these) is defined by the *intrinsic* or *acquired ability* of the mark to carve out the necessary *identity* and *distinction* for the source of the goods/services of one party from those of the others (*NOTE*: these are associations based on perceptions of customers).

Therefore, the "distinctiveness" of a mark is not quantified by any "absolute" standard, but rather judged on a "relative" scale. In this regard, Figure 5.1 will be helpful to see how marks can be distinguished on a relative scale.

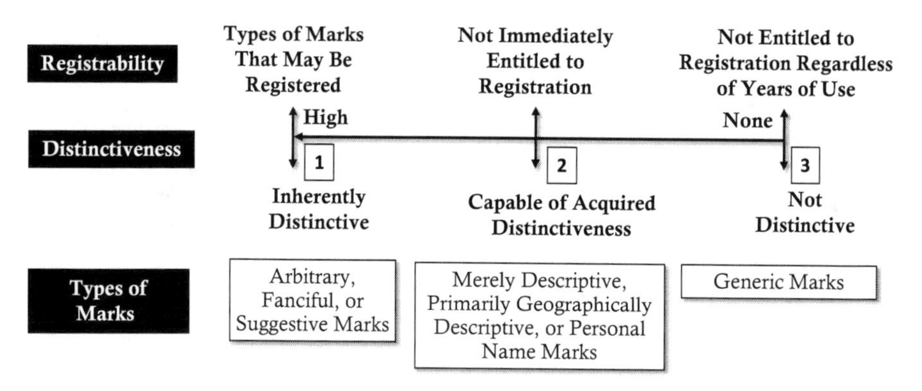

FIGURE 5.1
The distinctiveness continuum—marks and their registrability.

Thus, on the basis of the "distinctiveness" of marks (Figure 5.1), *trademarks* or *service marks* can be categorized as:

1. Those that are "inherently distinctive";
2. Those that are "capable of acquiring distinctiveness;" and
3. Those that are "not distinctive".

A general rule of thumb is: *the higher the distinctiveness of a mark, the easier it is for the owner to register the mark with the USPTO.* The following examples for marks will clarify these ideas better.

Marks That Are Inherently Distinctive

Inherently distinctive marks[12a] are marks that are *intrinsically capable* of signaling the source of goods/services of a party to the customer.[12b] Indeed, they are directly registrable with the USPTO. There are three such marks, namely—*arbitrary, fanciful,* and *suggestive* marks. Figure 5.1 depicts these and other kinds of marks.

1. **Arbitrary marks**. An "arbitrary mark" is a *trademark/service mark* based on a commonly used *word* or *symbol that is arbitrarily applied to a good or service.* In fact, these symbols or words *neither suggest a rational relationship nor describe the good/service* in any way.[13] Nevertheless, these marks serve as *strong marks* and function as source identifiers.

   The purpose of an *arbitrary mark* is *to build a unique and strong association* between *the mark* and *the product/service that the mark represents in the minds of consumers.* Thus, the **Apple** symbol is an arbitrary mark to represent a host of high technology products (*for which the symbol holds no relationship or meaning*), including—laptops, computers, iPhones, iPods, iPads, and other products, is a powerful example for this. Other examples are **Camel** for cigarettes, and **BlackBerry** for a type of cell phones;

2. **Fanciful marks**. A "fanciful mark" is a *trademark/service mark that is fancifully applied to a good or service*. Thus, like *arbitrary marks*, "fanciful marks" are invented and hold no connection or meaning to the products/services they represent.[14] They are totally made-up words, intended to be *inherently distinctive*.

   Examples of *fanciful marks* include—**Exxon** for an oil and gas corporation, **Pepsi** for a foods and beverages company, and **Rolex** for a company that makes watches; and

3. **Suggestive marks**. A "suggestive mark" is a *trademark/service mark* that *merely suggests some quality, characteristic, or attribute of the good or the service, but does not directly describe* it. Usually, an element of incongruity characterizes a *suggestive mark*. Typically, a *suggestive mark* requires *imagination or logical reasoning* to decipher the nature/quality of the goods or services that the mark represents.[15] In terms of distinctiveness, a *suggestive mark* ranks lower compared to an *arbitrary mark* or a *fanciful mark*, though it is still *inherently distinctive* (Figure 5.1). It is still protected without having to show any proof for secondary meaning.

   Examples of *suggestive marks* include—**Microsoft**, which is a linguistic blend of two words, **microco**mputers and **soft**ware, for an American multinational software company; and the **Starbucks** logo, based on an old sixteenth-century Norse woodcut, two-tailed mermaid, is *supposed to suggest that this brand of coffee would be as seductive as the mermaid herself*.[16] The **Mercedes-Benz** logo, a three-pronged star, is *supposed to suggest the automaker's drive toward universal motorization with its engines dominating the land, sea, and air* (three points).[17]

In this regard, the USPTO's exposition of *distinctiveness continuum* is instructive:[18]

> With regard to *trademark* significance, matter may be categorized along a continuum, ranging from *marks that are highly distinctive*, to *matter that is a generic name for the goods or services*. The degree of distinctiveness—or, on the other hand, descriptiveness—of a designation can be determined only by considering it in relation to the specific goods or services. *Remington Products, Inc. v. N. Am. Philips Corp.*, 892 F.2d 1576, 1580, 13 USPQ2d 1444, 1448 (Fed. Cir. 1990) (the mark must be considered in context, i.e., in connection with the goods).
>
> At one extreme are marks that, when used in relation to the goods or services, are *completely arbitrary or fanciful*. Next on the continuum are *suggestive marks*, followed by *merely descriptive matter*. Finally, *generic* terms for the goods or services are at the opposite end of the continuum from *arbitrary or fanciful marks*. As stated in *H. Marvin Ginn Corp. v. Int'l*

Ass'n of Fire Chiefs, Inc., 782 F.2d 987, 989, 228 USPQ 528, 530 (Fed. Cir. 1986), quoting *Weiss Noodle Co. v. Golden Cracknel & Specialty Co.*, 290 F.2d 845, 847, 129 USPQ 411, 413 (C.C.P.A. 1961): "[t]he name of a thing is in fact the ultimate in descriptiveness."

Fanciful, arbitrary, and suggestive marks, often referred to as *inherently distinctive* marks, are registrable on the Principal Register without proof of *acquired distinctiveness*. See TMEP §1209.01(a):

Marks that are *merely descriptive* of the goods or services may not be registered on the Principal Register absent a showing of *acquired distinctiveness* under 15 U.S.C. §1052(f). See TMEP §1209.01(b) regarding *merely descriptive marks*, and TMEP §§1212–1212.10 regarding acquired distinctiveness. *Merely descriptive marks* may be registrable on the Supplemental Register in applications under §1 or §44 of the Trademark Act 15 U.S.C. §1091. Matter that is *generic* for the goods or services is not registrable on either the Principal or the Supplemental Register under any circumstances. See TMEP §§1209.01(c)–(c)(iii).

Marks Capable of Acquired Distinctiveness

In some cases, a *trademark* or *service mark* may start out as a *descriptive mark* (*inherently non-distinctive*), but have the potential *to become distinctive*, by acquiring *secondary meaning* over time. One such famous example[19] is the *trademark*, **Coca-Cola**, which at the beginning was a *descriptive mark* (for a beverage, or cola, made from the ingredients of the coca plant) and hence, *inherently non-distinctive*. However, **Coca-Cola** today is one of the most instantly-recognized brand names around the world, because it has *acquired distinctiveness*. It must be remembered that *descriptive marks are not immediately entitled to registration*. Thus, the fact still remains, *descriptive marks* are unprotectable until they acquire *secondary meaning*.

There are three such *descriptive marks capable of acquired distinctiveness*, namely—*merely descriptive marks, primarily geographically descriptive marks*, and *personal name marks* (Figure 5.1).

1. **Merely descriptive marks**. According to 1209.01(b) TMEP:[18a]

> To be *refused registration* on the Principal Register under §2(e)(1) of the Trademark Act, 15 U.S.C. §1052(e)(1), a mark must be *merely descriptive* or *deceptively misdescriptive* of the goods or services to which it relates. A "merely descriptive mark" is a mark *that describes an ingredient, quality, characteristic, function, feature, purpose, or use of the specified goods or services.*[20]

Thus, *In re Bright-Crest, Ltd.*, 204 USPQ 591 (TTAB 1979) (**Coaster-Cards** held *merely descriptive* of a coaster suitable for direct mailing; *In re Bed & Breakfast Registry*, 791 F.2d 157, 229 USPQ 818 (Fed. Cir. 1986) (**Bed & Breakfast Registry** held *merely descriptive* of lodging reservations services); *In re Gyulay*, 820 F.2d 1216, 3 USPQ2d 1009 (Fed. Cir. 1987) (**Apple**

Pie held merely descriptive of potpourri).[20a] However, it is not necessary that a term describe all of the functions, characteristics, features, or purposes of a product to be considered *merely descriptive*; it is enough if it describes "one" significant attribute, function, or property. *In re Chamber of Commerce*, 675 F.3d at 1300, 102 USPQ2d at 1219.

2. **Primarily geographically descriptive marks.** A *geographically descriptive mark primarily identifies the place of origin* for the goods or services.

   The place of "origin" is defined as the *place of production of the goods* or the *location for rendering the services.* Clearly, the same *geographically descriptive mark* may be used by a wide variety of goods and services from the same place of origin, *which makes the term inherently non-distinctive.* This is the reason why these marks *are not immediately entitled to registration in the* USA.[20b] However, when such marks acquire "secondary meaning" (distinctiveness) over time, through use and promotion, they may qualify for *trademark protection.* They are also called as *geographical indications*, a separate class of IPRs in India, China, and the EU.

3. **Personal name marks.** These are also known as *primarily merely surnames.* Surnames are often shared by several individuals, and hence, are *not distinctive.* The U.S. Trademark Law recognizes the possibility *that each person may have an interest in doing business under his/her name.* Consequently, the USPTO demands a "long, substantial, and exclusive use" of the surname as a *trademark* or *service mark* before it can be registered.[21]

Marks That Are Not Distinctive

"Generic terms" *lack the source identifying ability.* Thus, when potential buyers understand that a "term" refers to *the general category of products or services*, and *not to the specific source of a product*, the term is "generic." Sometimes, a product/service may have more than one generic name (such as, elevator/lift; coin laundry/coin wash). Thus, when the public understands that there are two generic terms for a product/service, both the terms will be considered as "generic." It is not needed to establish whether the public uses these terms to refer to the relevant product/service. The right inquiry is whether the public understands that these terms are "generic." See, *In re 1800Mattress.com IP LLC*, 586 F.3d 1359, 92 USPQ2d 1682, 1685 (Fed. Cir. 2009). A key point to note is that, "generic marks" *are not entitled to registration regardless of years of use.*

According to the USPTO:[22]

> *Generic terms* are terms that the relevant purchasing public understands primarily as the common or class name for the goods or services. *In re*

> *Dial-A-Mattress Operating Corp.*, 240 F.3d 1341, 57 USPQ2d 1807, 1811 (Fed. Cir. 2001); *In re Am. Fertility Soc'y*, 188 F.3d 1341, 1346, 51 USPQ2d 1832, 1836 (Fed. Cir. 1999). These terms are *incapable of functioning as registrable trademarks denoting source and are not registrable* on the Principal Register under §2(f) or on the Supplemental Register.
>
> When a mark is comprised entirely of generic wording and some or all of the wording in the mark is the phonetic equivalent of the generic wording, the entire mark may not be disclaimed, even in the proper spelling, and approved for registration on the Supplemental Register. *The disclaimer does not render an otherwise unregistrable generic mark registrable.* See, TMEP §§1213.06 and 1213.08(c).

Unprotectable Marks

According to § 2 (15 U.S.C. § 1052),[23] there are several kinds of marks (*words, phrases, symbols, or designs, or combinations thereof*) that are *unprotectable* as marks for various reasons. Some of them are described below:

Immoral or Scandalous Marks

Section 2(a) of the Trademark Act, 15 U.S.C. §1052(a): *absolutely prohibits the trademark registration* of "immoral" or "scandalous" subject matter on either the Principal Register or Supplemental Registers. *Immoral* or *scandalous* marks are *shocking to the sense of truth, decency, or propriety; disgraceful; offensive; disreputable;* giving *offense to the conscience or moral feelings.* See, *In Re Fox*, 702 F.3d 633 (Fed. Cir. 2012).

In the case of *immoral* or *scandalous* marks, two criteria must be ascertained: first, the meaning evoked by a mark in the *context of contemporary attitudes, and relevant marketplace* for the goods or services outlined in the application and second, the meaning communicated by the mark, in terms of not necessarily what a majority thinks, but how *a sizable cross-section of the general public* perceives it. Therefore, an examining attorney may reject a proposed mark on the grounds that it is *immoral* or *scandalous,* only if he/she can show that: (a) *a sizable cross-section of the general public would consider the mark to be immoral* or *scandalous;* (b) in the *context of contemporary attitudes; and* (c) in the *relevant marketplace.*

According to the USPTO:[24]

> Although the words *immoral* and *scandalous* may have *somewhat different connotations,* case law has included *immoral matter* in the same category as *scandalous matter. See In re McGinley,* 660 F.2d 481, 484 n.6, 211 USPQ 668, 673 n.6 (C.C.P.A. 1981), *aff'g* 206 USPQ 753 (TTAB 1979) ("Because of

our holding, *infra*, that appellant's mark is *'scandalous,'* it is unnecessary to consider whether appellant's mark is *'immoral.'* We note the dearth of reported *trademark* decisions in which the term *'immoral'* has been directly applied.").

Two *hypothetical* examples of *immoral marks* could be—**Hindu Sacred Beef** for beef products and **Sin Paradise** for a highway hotel.

Similarly, two *hypothetical* examples of *scandalous marks* could be—**Gandhiji Gun Store** for a gun shop and **Pope's Hide Away** for a vacation resort.

Deceptive Marks

Section 2(a) of the Trademark Act, 15 U.S.C. §1052(a), *absolutely prohibits the trademark registration* of "deceptive" subject matter on either the Principal Register or Supplemental Registers. Neither a disclaimer of the deceptive matter nor a claim that it has acquired distinctiveness under §2(f) can obviate a refusal on the ground that the mark consists of or comprises *deceptive matter.* See, *Am. Speech-Language-Hearing Ass'n v. Nat'l Hearing Aid Society,* 224 USPQ 798, 808 (TTAB 1984); *In re Charles S. Loeb Pipes, Inc.*, 190 USPQ 238, 241 (TTAB 1975).[24] "Deceptive marks" may include: (a) those that *falsely describe the material content of a product*[25] or (b) those that are *geographically deceptive.*[26]

Marks that falsely describe the material content of a product. It is important to know that any mark containing a term identifying a material, ingredient, or feature cannot be refused registration under §2(a), if the mark in its entirety *would not be perceived* as indicating that the goods contained that material or ingredient. For example, the mark **COPY CALF** is "not" considered "deceptive" for wallets and billfolds that are made from synthetic/plastic materials *that imitate leather,* because the expression "copy cat" suggested to purchasers that *the goods were imitation items made of calf skin.*[27] On the other hand, **TEXHYDE**, as applied for *synthetic fabric*, and **SOFTHIDE**, as applied for *imitation leather*—were held "deceptive" because these marks fit the description of *deceptive marks.*[28]

Similarly, the term **GOLDEN** may "not" be considered "deceptive," while the term **GOLD** would be considered "deceptive" *for jewelry not made of gold.* In addition, formatives and other grammatical variations of a term may "not" necessarily be "deceptive" in relation to the relevant goods. For instance, the term "silky" normally means, "resembling silk."[29] Thus, a mark containing the term **SILKY** may "not" be considered "deceptive" [though it might not be registrable under §2(e)(1)]. Thus, it is a good practice to carefully review the dictionary definitions of such terms to determine the what the term might mean to future buyers.

According to USPTO:[30]

> A *deceptive mark* may be comprised of: (1) a single *deceptive* term; (2) a *deceptive* term embedded in a *composite mark* that includes additional non-deceptive wording and/or design elements; (3) a *term or a portion of a term* that alludes to *a deceptive quality, characteristic, function, composition*, or use (*see Am. Speech-Language-Hearing Ass'n v. Nat'l Hearing Aid Society*, 224 USPQ 798, 808 (TTAB 1984)]; (4) the phonetic equivalent of a *deceptive* term [*see In re Organik Technologies, Inc.*, 41 USPQ2d 1690, 1694 (TTAB 1997); *Tanners' Council of Am., Inc. v. Samsonite Corp.*, 204 USPQ 150, 154 (TTAB 1979)]; or (5) the foreign equivalent of any of the above (see, e.g., *Palm Bay Imps., v. Veuve Clicquot Ponsardin Maison Fondee En 1772*, 396 F.3d 1369, 1377, 73 USPQ2d 1689, 1696 (Fed. Cir. 2005)]. Although there is no published Board or Federal Circuit decision regarding whether a mark consisting solely of a design can be deceptive, if there is evidence to support such a refusal, it should be issued.

Geographically deceptive marks. Section 2(a) of the Trademark Act, 15 U.S.C. §1052(a), *prohibits* the registration of a *geographical indication* which—when used on or in connection with wines or spirits—*identifies a place other than the origin of the goods.*[31a]

According to TMEP 1210.01:[31b]

> To support a refusal of registration on the grounds that a *geographic term* is *deceptive* under §2(a), the examining attorney must show that:

1. The primary significance of the mark is a generally known geographic location [see TMEP §§1210.02–1210.02(b)(iv)];
2. The goods or services do not originate in the place named in the mark (see TMEP §1210.03);
3. Purchasers would be likely to believe that the goods or services originate in the geographic place identified in the mark [see TMEP §§1210.04–1210.04(d)]. Note: If the mark is remote or obscure, the public is unlikely to make a goods/place or services/place association [see TMEP §1210.04(c)]; and
4. The misrepresentation is a material factor in a significant portion of the relevant consumer's decision to buy the goods or use the services.

Disparaging Marks

Section 2(a) of the Trademark Act, 15 U.S.C. §1052(a): *prohibits the registration of a mark* (on either the Principal or the Supplemental Register) *that comprises or consists of matter which does any of the following*, to persons, institutions, beliefs, or national symbols: (a) *disparaging them;*[32a] (b) *bringing them into contempt* and *disrepute;*[32a] or (c) *falsely suggesting a connection to them.*[32b]

See, Section 5.1 above for the definition of a "person" in the U.S. Trademark Law. Further, Section 2(a) of the Trademark Act, 15 U.S.C. §1052(a): protects, *inter alia*, "persons, living or dead." According to the USPTO: a "national symbol" is "subject matter of unique and special significance that, because of its meaning, appearance, and/or sound, immediately suggests or refers to the country for which it stands. In the case, *In re Consol. Foods Corp.*, 187 USPQ 63, 64 (TTAB 1975) (noting that the U.S. national symbols include the *bald eagle, Statue of Liberty, American flag, Presidential symbol, designation Uncle Sam and the unique human representation thereof,* and the *heraldry and shield designs used in governmental offices*). National symbols include the symbols of foreign countries, as well as those of the United States. In the case, *In re Anti-Communist World Freedom Cong., Inc.*, 161 USPQ 304, 305 (TTAB 1969)."[33] It is important to note that the U.S. Trademark Law *does not bar the registration of marks comprising national symbols; it only prohibits* the registration of marks that either *disparage national symbols* and/*suggest a false connection* or *denigrate them to contempt or disrepute.*[33]

According to the USPTO:[34,35]

> The Board applies *a two-part test* in determining *whether a proposed mark is disparaging*:

a. What is the likely meaning of the matter in question, taking into account not only dictionary definitions, but also the relationship of the matter to the other elements in the mark, the nature of the goods or services, and the manner in which the mark is used in the marketplace in connection with the goods or services; and
b. If that meaning is found to refer to identifiable persons, institutions, beliefs, or national symbols, whether that meaning may be disparaging to a substantial composite of the referenced group.

Some examples would be helpful to clarify the points discussed above:

- ***Disparaging***. In the case, *In re Lebanese Arak Corp.*, 94 USPQ2d 1215 (TTAB 2010),[36] **KHORAN** *for wines* was found to be "disparaging" because *the Muslim Americans*, in particular (and the others, in general), *would regard the mark* **KHORAN** *as referring to the holy text of Islam*; further, given that alcohol is a prohibited substance according to the Islamic religion, Muslims would find **KHORAN** used for wine as "disparaging" *to themselves, their religion, and their beliefs*; similarly, in *In re Heeb Media LLC*, 89 USPQ2d 1071 (TTAB 2008),[37] **HEEB** *for clothing and entertainment services* was found to be "disparaging" because of multiple reasons: **HEEB** means *a Jewish person*, dictionary definitions unanimously support the *derogatory nature of* **HEEB**, evidence of record supports that a substantial composite of the referenced group, for example, *the Jewish community*, will perceive **HEEB** as "disparaging," and **HEEB** *has no other meaning in relation to clothing or entertainment services.*

- ***Bringing them into contempt and disrepute.*** In the case, *In re Anti-Communist World Freedom Cong., Inc.*, 161 USPQ 304 (TTAB 1969), the design of an "X" superimposed over a hammer and sickle, a national symbol of the Union of Soviet Socialist Republics (U.S.S.R.), was found to be "disparaging" and "bringing in contempt and disrepute." Similarly, in the case, *Greyhound Corp. v. Both Worlds Inc.*, 6 USPQ2d 1635, 1639-40 (TTAB 1988),[38] the design of dog defecating, for clothing, was found to be "disparaging" and "bringing in contempt and disrepute" to the Greyhound Corporation's running dog symbol; the Board also found evidence of record "sufficient to show prima facie that this design (the running dog symbol) is, in effect, an alter ego of opposer which points uniquely and unmistakably to opposer's persona."
- ***Suggesting a false connection.*** To assess that a mark *suggests a false connection to a person or an institution*, one must establish that: (1) the proposed mark is *either the same* or *a close approximation of the name/identity formerly used by another person or institution*; (2) the mark *uniquely and unmistakably points out to that person or institution*; (3) the *person* or *institution* named by the mark *is not connected with the activities performed by the applicant under the mark*; and (4) when the mark is used with the applicant's goods or services, *a connection with the person or institution would be automatically assumed*, because of the fame or reputation of the person or institution.

  In the case, *In re Shinnecock Smoke Shop*, 571 F.3d 1171, 91 USPQ2d 1218 (Fed. Cir. 2009) the SHINNECOCK BRAND FULL FLAVOR and SHINNECOCK BRAND LIGHTS, both for cigarettes, *were found to falsely suggest a connection with the Shinnecock Indian Nation* (person); on the contrary, in the case, *Univ. of Notre Dame du Lac v. J.C. Gourmet Food Imps. Co.*, 703 F.2d 1372, 1377, 217 USPQ 505, 509 (Fed. Cir. 1983), *aff'g* 213 USPQ 594 (TTAB 1982), "Notre Dame and design for cheese," were found to "not falsely suggest a connection with the University of Notre Dame (institution)." In addition, the Board noted that, "Notre Dame" *cannot be considered as a name just associated with the University*, as it *serves to identify a famous and sacred religious figure and is used in the names of churches dedicated to Notre Dame*, such as the Cathedral of Notre Dame in Paris, France.

Marks Likely to Cause Confusion with an Existing Mark

Section 2(d) of the Trademark Act, 15 U.S.C. §1052(d),[39] *bars the registration of a mark—which so resembles a mark already registered with* the USPTO, or *a mark/trade name previously used in the United States by another and not abandoned*—likely *to cause confusion, mistake,* or *deception,* when used on/in connection with the goods/services of the applicant. Section 2(d) applies regardless of whether the applicant sought registration of the mark on the Principal Register or the Supplemental Register.

The examining attorney must conduct a search of the USPTO records in order to determine whether the applicant's mark so resembles any registered mark(s) *such that it is likely to cause confusion* or *mistake,* when used on/in connection with the goods/services identified in the application. Thus, if the examining attorney *determines that there is a likelihood of confusion between applicant's mark and a previously registered mark or marks,* then he/she will refuse the registration of the applicant's mark under §2(d).

Unprotectable Functional Matter

According to the U.S. Trademark Law: "functional matter" cannot be protected as a *trademark.* 15 U.S.C. §§1052(e)(5) and (f), 1064(3), 1091(c), and 1115(b).[40] A feature is *functional,* "if it is essential to the use or purpose of the article or if it affects the cost or quality of the article." *TrafFix Devices, Inc. v. Mktg. Displays, Inc.,* 532 U.S. 23, 33, 58 USPQ2d 1001, 1006 (2001);[41a] *Qualitex Co. v. Jacobson Prods. Co.,* 514 U.S. 159, 165, 34 USPQ2d 1161, 1163-64 (1995);[41b] *Inwood Labs., Inc. v. Ives Labs., Inc.,* 456 U.S. 844, 850, n.10, 214 USPQ 1, 4, n.10 (1982).[41c] If it is determined that *a proposed mark is functional,* for public policy reasons, *then it would be absolutely barred from registration* on either the Principal or the Supplemental Register, even when there is evidence to show that the proposed mark has acquired distinctiveness. Generally, the "functionality" is established on the basis one or more of the following factors, known to as the "*Morton-Norwich* factors":[42]

- The existence of a utility patent that discloses the utilitarian advantages of the design sought to be registered;
- Advertising by the applicant that touts the utilitarian advantages of the design;
- Facts pertaining to the availability of alternative designs; and
- Facts pertaining to whether the design results from a comparatively simple or inexpensive method of manufacture.

In re Becton, Dickinson & Co., 675 F.3d 1368, 1374-75, 102 USPQ2d 1372, 1377 (Fed. Cir. 2012); *In re Morton-Norwich Prods., Inc.,* 671 F.2d 1332, 1340–1341, 213 USPQ 9, 15–16 (C.C.P.A. 1982).

Primarily Geographically Deceptively Misdescriptive Marks

Section 2(e)(1) of the Trademark Act, 15 U.S.C. §1052(e)(1), prohibits the registration of marks that are "deceptively misdescriptive" of the goods/services to which they are applied. However, the examining attorney must consider the mark in relation to the applicant's goods/services to determine whether a mark is "deceptively misdescriptive." As discussed above, *a term that conveys an immediate idea of an ingredient, quality, characteristic, function, or featureT of the goods/services with which it is used* is "merely descriptive."[43] If a term *immediately conveys such an idea, but the idea itself is false,* although plausible, *then the term* is "deceptively misdescriptive" and *is unregistrable under* §2(e)(1). See, *In re Woodward & Lothrop Inc.*, 4 USPQ2d 1412 (TTAB 1987) (CAMEO "deceptively misdescriptive" of jewelry); *In re Ox-Yoke Originals, Inc.*, 222 USPQ 352 (TTAB 1983) (G.I. "deceptively misdescriptive" of *gun cleaning patches, rods, brushes, solvents, and oils*).

The Trademark Act does not prohibit the registration of *misdescriptive terms* unless they are "deceptively misdescriptive," that is, unless the persons who encounter the mark, as used on or in connection with the goods/services in question, *are likely to believe the misrepresentation.* See, *Binney & Smith Inc. v. Magic Marker Indus., Inc.*, 222 USPQ 1003 (TTAB 1984) (LIQUID CRAYON *did not* hold a *common descriptive name*, or *merely descriptive*, or "deceptively misdescriptive," of coloring kits or markers).

In order to reject the registration of a mark that is "primarily geographically deceptively misdescriptive," the examining attorney must show that:

- The primary significance of the mark is *a generally known geographic location;*[44a]
- The goods/services *do not originate in the place identified in the mark;*[44b]
- Purchasers *would be likely to believe that the goods or services originate in the geographic place identified in the* mark.[44c] *Note:* If the mark is remote or obscure, the public is unlikely to make a goods/place or services/place association[44d]; and
- The misrepresentation is *a material factor in a significant portion of the relevant consumer's decision to buy the goods or use the services.*[44e]

In re Spirits Int'l, N.V., 563 F.3d 1347, 1350–1354, 90 USPQ2d 1489, 1490–1495 (Fed. Cir. 2009); *In re Les Halles De Paris J.V.*, 334 F.3d 1371, 67 USPQ2d 1539 (Fed. Cir. 2003); *In re Cal. Innovations Inc.*, 329 F.3d 1334, 66 USPQ2d 1853 (Fed. Cir. 2003), *reh'g denied*, 2003 U.S. App. LEXIS 18883 (Fed. Cir. Aug. 20, 2003).

Finally, according to the WIPO, the following requirements serve as *grounds for refusal* of a *trademark* or *service mark*:[44f]

> Generally speaking, *two different kinds of requirements* can be distinguished. The *first* relates to the basic function of a *trademark*, namely, *its function to distinguish the products or services of one enterprise from the products or services of other enterprises.* From that function it follows *that a trademark must be distinguishable.* The *second* kind of requirement relates to *the potential negative effects of a trademark if it is misleading or if it is contrary to public order or morality.*

These two kinds of requirements exist in practically all national *trademark* laws. They also appear in Article 6*quinquies* (B) of the Paris Convention for the Protection of Industrial Property (Paris Convention) (for the full text of Article 6*quinquies*, see the Annex II), which states that *trademarks* enjoying protection under Article 6*quinquies* (A) may be denied registration only if *they are devoid of any distinctive character* or if *they are contrary to morality or public order and, in particular, of such a nature as to deceive the public.*

Other Important Information Concerning Trademarks

1. **Duration of the term.**[45] Unlike patents or copyrights, the *trademarks last indefinitely* as long as the *trademark is in real use,* to identify the goods or services. The *normal term* of a U.S. trademark is 10 years, with 10-year *renewal terms.* However, in order to maintain registration, the USPTO requires the *trademark* owner to file a "Declaration of Use" affidavit between the 5th and 6th year, following the initial registration. Nevertheless, the USPTO *does not remind* the *trademark* owner of this deadline. If an affidavit is NOT filed, however, the *registration will simply be cancelled* and *cannot be reinstated.* Therefore, owners wishing to retain the *trademark* must diligently apply for renewal of registration within the last year of each term.

   The USPTO *allows a grace period of 6 months for renewal,* for an additional fee.
2. **Loss of trademark rights.**[45] Firstly, it must be remembered that if the use of a *trademark* in commerce is discontinued, the mark will be considered abandoned. Thus, *trademarks* are protected *only when they are in continual use,* regardless of valid registration and diligent renewal. Further, *trademark* rights may also be lost if the mark has been *improperly licensed.* Finally, a *distinctive trademark* may become *generic* over time, *resulting in loss of rights.* For example, "aspirin," the *distinctive trademark* of the Bayer company, has become *generic* over time, and is no longer subject to *trademark* protection, in the USA, but remains under *trademark* protection in Canada and in many European countries.

International Treaties Administered by WIPO

The following international treaties[46] on that WIPO administers, in combination with the national and regional laws, make up the international legal framework for *trademarks:*[47] (1) Paris Convention; (2) Madrid Agreement and

Madrid Protocol; (3) Nice Agreement; (4) Vienna Agreement; (5) Trademark Law Treaty; and (6) Singapore Treaty.

A brief description of these treaties are as follows:

1. **Paris Convention.**[1] The "Paris Convention" for the protection of industrial property was adopted in 1883. This is the first international treaty between many nations *to provide protection to the intellectual works of all the states that are signatories to this treaty*. It applies to *trademarks, service marks, trade names* (designations under which industrial or commercial activity is carried out), among other forms of industrial property;
2. **Madrid Agreement and Madrid Protocol.**[47a] The "Madrid System for the International Registration of Marks" is governed by two agreements: (a) the *Madrid Agreement*, that concluded in 1891 and (b) the *Madrid Protocol*, that concluded in 1989. The *Madrid Agreement* and the *Madrid Protocol* are parallel and independent treaties which are open to the contracting parties of the Paris Convention. This system enables the protection of a *trademark/service mark* possible in all the contracting parties, by obtaining *a single international registration for trademark/service mark*, by simply filing one application with the International Bureau (through the office of the home country), in one language (English, French, or Spanish), and paying one set of fees.
3. **Nice Agreement.**[47b] The "Nice Agreement" that concluded in 1957 at Nice, revised in 1967 at Stockholm (*the Stockholm Act*), and 1977 at Geneva (*the Geneva Act*), and amended in 1979—*establishes a classification of goods and services* for the purposes of registering *trademarks* and *service marks* (the Nice Classification). The agreement is open to all states that are party to the Paris Convention.
4. **Vienna Agreement.**[47c] The "Vienna Agreement" that concluded in 1973 at Vienna, and amended in 1985, establishes *a classification for marks* (known as *the Vienna Classification*) that consist of/contain, figurative elements, and is open to all signatories of the Paris Convention.
5. **Trademark Law Treaty**. The "Trademark Law Treaty" *standardizes and streamlines the procedures of national and regional trademark registration.*

   According to WIPO:[47d]

   > This is achieved through the simplification and harmonization of certain features of those procedures, thus making *trademark* applications and the administration of *trademark* registrations in multiple jurisdictions less complex and more predictable. The great majority of the provisions of the Trademark Law Treaty concern the procedure before a *trademark*

---

[1] See chapter 2 for full details.

office which can be divided into three main phases: (1) application for registration; (2) changes after registration; and (3) renewal. The rules concerning each phase are constructed so as to clearly define the requirements for an application or a specific request.

6. **Singapore Treaty.**[47e] The "Singapore Treaty" is a *modern and dynamic international framework for the harmonization of administrative trademark registration procedures.* It is built on the foundation of the *Trademark Law Treaty of 1994* described above. In addition to having a broader scope of application, the *Singapore Treaty* includes more recent developments in the field of communication technologies. Thus, the *Singapore Treaty* is applicable to all types of marks registrable under the law of a given contracting party and allows a contracting party to use an appropriate means of communication with their offices (in electronic form or by electronic means of transmittal).

Infringement of a Trademark/Service Mark

"Trademark infringement" is the *unauthorized use* of a *trademark* or *service mark on* or *in connection with goods and/or services,* in a manner *that is likely to cause confusion, deception,* or *mistake about the source of the goods and/or services.* In this regard, the *Lanham Act* provides other related causes of action, including imposing liability for: (a) *unauthorized use* of a *trademark* or *service mark;* (b) *false designation of origin* that may cause confusion among the relevant consumer population; (c) *false advertising* that may result in *trademark* dilution or commercial loss, and/or competitive damage; and (d) *bad faith registration, trafficking, or use of an Internet domain name* that is identical to or imitation of a real *trademark* or *service mark.*

15 U.S.C. § 1114(1)[48] defines the "infringement of a *trademark/service mark,*" as:

Any person who shall, *without the consent of the registrant:* -

a. *Use in commerce* any reproduction, counterfeit, copy, or colorable imitation of a registered mark *in connection with the sale, offering for sale, distribution, or advertising* of *any goods or services* on or in connection with which *such use is likely to cause confusion, or to cause mistake, or to deceive;* or
b. *Reproduce, counterfeit, copy, or colorably imitate* a registered mark and apply such reproduction, counterfeit, copy, or colorable imitation to labels, signs, prints, packages, wrappers, receptacles, or advertisements intended *to be used in commerce* upon or *in connection with the sale, offering for sale, distribution, or advertising of goods or services*

> on or in connection with which *such use is likely to cause confusion, or to cause mistake, or to deceive,*
>
> *shall be liable in a civil action* by the registrant for the remedies hereinafter provided. Under subsection (b) hereof, the registrant shall not be entitled to recover profits or damages *unless the acts have been committed with knowledge that such imitation is intended to be used to cause confusion,* or *to cause mistake,* or *to deceive.*

Clearly, the central issue here is the "likelihood of confusion." To establish *infringement* under the U.S. Trademark Law for either a *registered mark* under 15 U.S.C. § 1114, or *an unregistered mark* under 15 U.S.C. § 1125(a), the plaintiff must demonstrate that: (1) the mark is *valid and legally protectable;* (2) the *mark is owned;* and (3) the defendant's *use of the mark* to identify goods or services *causes the likelihood of confusion, mistake, or deception.* See, *A&H Sportswear, Inc. v. Victoria's Secret Stores, Inc.,* 237 F.3d 198 (3rd Cir. 2000).[49a]

Further, to support a *trademark infringement* claim in court, the plaintiff must establish that: (1) it has *a valid mark entitled to protection;* (2) that the *defendant used the same or a similar mark* (3) *in commerce* (4) *in connection with the sale or advertising of goods or services;* and (5) without the plaintiff's consent. In addition, the plaintiff must show that (6) the defendant's *unauthorized use of the mark* (7) *is likely to cause confusion* as to the affiliation, connection, or association of defendant with plaintiff, or as to *the origin, sponsorship, or approval of defendant's goods, services, or commercial activities by plaintiff.* See, *1-800 Contacts, Inc. v. WhenU.com, Inc.,* 414 F.3d 400 (2d Cir. 2005).[49b] Thus, three distinct elements, namely—"use," "in commerce," and "likelihood of confusion"—are necessary to establish a *trademark* infringement claim.

Generally, the courts consider a variety of factors (such as shown below) *in order to establish whether there is a likelihood of confusion among consumers:*[49c]

- The *strength of the plaintiff's mark;*
- The *degree of similarity* between the marks at issue;
- Whether or not *the parties' goods and/or services are sufficiently related;*
- The *defendant's intent* in adopting its mark;
- The *range of prospective purchasers* of the goods or services;
- Where and how are the parties' *goods or services advertised, marketed, and sold;*
- The *purchasing conditions;*
- Whether or not *the consumers are likely to assume* (mistakenly) *that they come from a common source;* and
- Whether or not there is *real evidence of actual confusion caused by the allegedly infringing mark.*

Other Causes of Action Related to Trademarks/Service Marks

The Lanham Act provides a set of valuable tools for *claims of unfair competition* against a person's goods, services or commercial activities, specifically—*false designation of origin* and *false advertising*. They will be described below:

According to 15 U.S.C. § 1125(1):[50]

> Any person who, *on/in* connection with any goods or services, or any container for goods, *uses in commerce* any word, term, name, symbol, or device, or any combination thereof, or *any false designation of origin, false or misleading description of fact*, or *false or misleading representation of fact*, which—

a. *Is likely to cause confusion*, or *to cause mistake*, or *to deceive* as to the affiliation, connection, or association of such person with another person, or *as to the origin, sponsorship, or approval* of his or her goods, services, or commercial activities by another person; or
b. *In commercial advertising or promotion, misrepresents* the nature, characteristics, qualities, or geographic origin of his or her or another person's goods, services, or commercial activities,

   shall be liable in a civil action by any person who believes that he or she is or is likely to be damaged by such act.

False Designation of Origin

As shown above, 15 U.S.C. § 1125(1)(a) provides a cause of action for use of *any mark* or *false designation of origin* that is likely to cause confusion, or to cause mistake, or to deceive regarding the "origin, sponsorship, or approval of goods, services or commercial activities." There are two important ramifications to this:

Infringement of an unregistered trademark. Most countries of the world (except the United States of America), including the European nations—follow a *first-to-file rule* for *trademark* rights, *which necessitates the registration of a trademark*. However, the USA follows a *first-to-use-rule* where *trademark* rights *result from the use of the mark in commerce*. Indeed, this is valuable for *claims of unfair competition* in the case of an unregistered *trademark*.

According to some legal scholars:[51a]

> Where *two parties employ the same mark*—or *confusingly similar marks*—on related goods, "the exclusive right to the use of a trademark is founded

on priority of appropriation" (*Hanover Star Milling Co v Metcalf*, 240 US 403, 415 (1916); see also *United Drug Co v Theodore Rectanus Co*, 248 US 90, 97 (1918): "[T]he right to a particular mark *grows out of its use, not its mere adoption*"). Although federal registration provides several strategic benefits, *the first-use system renders registration unnecessary to establish rights.* As such, it is crucial for brand owners *to beware of the rights wielded by unregistered trademarks in the United States, which are protected under both common law and the federal Lanham Act.*

It is important to note that although the common law and the Lanham Act provide protection against infringement for "unregistered," but "in-use" *trademarks/service marks* in the United States, federal registration of *trademarks/service marks* is highly recommended, as it is very helpful in several circumstances.[51b]

The negative impact of trade names and other non-mark indications of source. Sometimes, it is possible that someone may use a *Trade Name* (not a *trademark*) or a *non-mark symbol* for goods, services, or commercial activities by another person and *create confusion in the minds of customers.* The Lanham Act provides a cause of action under such circumstances.

False Advertising

As shown above, 15 U.S.C. § 1125(1)(b) provides a cause of action for *misrepresenting* the nature, characteristics, qualities, or geographic origin of his/her or another person's goods/services/commercial activities, *in commercial advertising or promotion.* Notable is the fact that courts *do not require proof of literal falsification* and *it is sufficient if there is explicit or implicit misrepresentation.*

An example may illustrate the points better. Let us assume that there are two competitor companies, A & B, that sell pain relievers. *False advertising* can take place under four different scenarios, through advertisement:

Scenario-1. Company B *falsely accuses* (without proof) that Company A's pain reliever *upsets stomach.*

Scenario-2. Company B *falsely claims* (without proof) that its own pain reliever is *the best in the market, as it does not upset stomach.*

Scenario-3. Company B *falsely claims* (without proof) that *its own pain reliver works instantly, while Company A's product takes several hours to deliver pain relief.*

Scenario-4. Company B *falsely depicts* (without proof) the *commercial activities* of Company A, *stating that the latter discriminates the LGBT community.*

Bad Faith Registration and Cybersquatting

Bad Faith Registration or Fraud

In the USA, "bad faith" registration of *trademarks/service marks* is referred to as "fraud." According to the USPTO:[52,53] "Improper use of the federal registration symbol *that is deliberate and intended to deceive or mislead the public* is *fraud;*" also "Improper use of the federal registration symbol, ®, *that is deliberate and intends to deceive or mislead the public* or the USPTO is *fraud.*" See, *Copelands' Enterprises Inc. v. CNV Inc.*, 945 F.2d 1563, 20 USPQ2d 1295 (Fed. Cir. 1991); *Wells Fargo & Co. v. Lundeen & Associates*, 20 USPQ2d 1156 (TTAB 1991). However, an analysis of the use of federal registration symbols shows that *misunderstandings are more common than occurrences of actual fraudulent intent.* Common *reasons for improper use* of the federal registration symbol *that do not indicate fraud* are:[52]

- Mistaken understanding of the requirements of giving notice (confusion between *the notice of trademark registration*, which may not be given until after registration, and *notice of claim of copyright*, which must be given before publication by placing the notice © on material when it is first published);
- Carelessness in giving appropriate instructions to the printer, or misunderstanding, or voluntary action by the printer;
- The mistaken belief that registration in a state or foreign country gives a right to use the registration symbol [see *Brown Shoe Co., Inc. v. Robbins*, 90 USPQ2d 1752 (TTAB 2009); *Du-Dad Lure Co. v. Creme Lure Co.*, 143 USPQ 358 (TTAB 1964)];
- Registration of a partial-mark [see *Coca-Cola Co. v. Victor Syrup Corp.*, 218 F.2d 596, 104 USPQ 275 (C.C.P.A. 1954)];
- Incorrect mark registration for other goods [see *Duffy-Mott Co., Inc. v. Cumberland Packing Co.*, 424 F.2d 1095, 165 USPQ 422 (C.C.P.A. 1970), *aff'g* 154 USPQ 498 (TTAB 1967); *Meditron Co. v. Meditronic, Inc.*, 137 USPQ 157 (TTAB 1963)];
- A cancelled registration or recent expiry of the subject mark [see *Rieser Co., Inc. v. Munsingwear, Inc.*, 128 USPQ 452 (TTAB 1961)]; and
- Another mark to which the symbol relates on the same label (see *S.C. Johnson & Son, Inc. v. Gold Seal Co.*, 90 USPQ 373 (Comm'r Pats. 1951)].

Cybersquatting—Bad Faith Registration, Trafficking, and Use of an Internet Domain Name[54a, b]

"Cybersquatting" is a type of *cybercrime*, that may be defined as: *the bad faith registering, sale or use of a domain name, specifically with the intent of capitalizing on the goodwill of someone else's mark and profiting from it.* Thus, *cybersquatting* involves *three criminal offenses*: (a) bad faith registering; (b) intentionally stealing/misspelling a domain name, to significantly increase the website visits; and (c) directly profiting from it. In situations where *trademark* or *copyright owners* neglect to reregister their domain names, *cybersquatters*, who are actively looking for easy opportunities, steal domain names. *Cybersquatting* also includes advertisers who mimic the domain names of popular, highly visited websites. *Cybersquatting*, sometimes referred to as *domain squatting*, is one of several types of cybercrimes.

An example could be helpful here.[54a] Thus, in 2007, Dell filed a lawsuit against a party that registered the URL, *DellFinacncialServices.com*, and a very large number of other domain name owners, for *cybersquatting*. In this particular case, the defendants engaged in *bad faith registration of misspelled and confusingly similar domains to those owned by Dell*, with *the intention of capturing the traffic* from people mistyping *Dell Financial Services.com* and *profiting from it.*

Cybersquatters depend upon the goodwill of someone else's powerful *trademark*. By buying up domain names linked to a pre-existing business or person, *cybersquatters hope to profit through an association with well-known trademarks* or through *sale of the domain to the trademark owner*. In addition, *cybersquatters* sometimes engage in even more nefarious pursuits, such as *identity theft* of unsuspecting users, mistyping the URL.[54b] Fortunately, legal remedies are available in the USA for recovering from cybersquatting-related crimes, including the *Anti-Cybersquatting Consumer Protection Act*.[55] The *Anti-Cybersquatting Consumer Protection Act* enables a *trademark* owner to sue an alleged *cybersquatter* in a U.S. Federal court. If the plaintiff (*trademark* owner) wins the lawsuit, the Federal court may issue a court order requiring the defendant (cybersquatter) to transfer the domain name to the plaintiff and, when appropriate, pay monetary damages as well. Further, the *Internet Corporation for Assigned Names and Numbers*—a nonprofit organization that is charged with oversight responsibilities for domain name registration—utilizes rigorous standards for acceptance of domain names,[56] so that *assignment of a domain name is done with great diligence*. In addition, the *Internet Corporation for Assigned Names and Numbers* urges *trademark* owners *to be more responsible and renew their registrations on an annual basis* and *to report any cases of misuse to the agency as soon possible*. Finally, *Internet Corporation for Assigned Names and Numbers has built demanding requirements for domain name recovery.*

References

1. (a) https://www.uspto.gov/sites/default/files/documents/BasicFacts.pdf; (b) According to the WIPO website, *A trademark is a sign capable of distinguishing the goods or services of one enterprise from those of other enterprises. Trademarks are protected by intellectual property rights.* See, http://www.wipo.int/trademarks/en/;

and (c) All the countries that are party to the *Nice Agreement* constitute a Special Union within the framework of the *Paris Union for the Protection of Industrial Property*. They have adopted *a common classification of goods and services*, known as the *Nice Classification*, for the purposes of the registration of marks. Therefore, *trademarks* are *service marks* share the same classification system. See, http://www.wipo.int/export/sites/www/classifications/nice/en/pdf/8_list_class_order.pdf
2. (a) The *Lanham Act* (or the *Trademark Act* of 1946) is the *federal statute* governing *trademarks, service marks*, and *unfair competition*, signed into law by President Harry Truman. The *Lanham Act* (15 U.S.C. §§ 1051–1127) took effect on July 5, 1947. Further, it is the *Lanham Act* that grants the *administrative authority* to the USPTO for the trademark registration. Thus, the U.S. Trademark Law is mainly governed by the *Lanham Act, although State law continues to add its own protection, complementing (and often complicating) the federal trademark system*. Recently, the U.S. Trademark Law included the *Federal Trademark Dilution Act of 1995*, the 1999 *Anticybersquatting Consumer Protection Act*, and the *Trademark Dilution Revision Act of* 2006. See, https://www.uspto.gov/sites/default/files/trademarks/law/Trademark_Statutes.pdf; and (b) See Reference 2(a), pages 41–43.
3. https://www.google.com/search?rlz=1C1CHBD_enUS756US756&tbm=isch&q=Company+Logos+and+Brands&chips=q:company+logos+and+brands, g_1:symbol&sa=X&ved=0ahUKEwjQyerl3sbYAhUDRiYKHZBBDdcQ4lYIKSgB&biw=1824&bih=1043&dpr=1.5 mm
4. (a) http://www.thelogofactory.com/top-100-brand-logos/; and (b) https://euipo.europa.eu/ohimportal/en/trade-mark-definition
5. https://www.qualitylogoproducts.com/promo-university/top-10-advertising-jingles.htm
6. (a) http://www.wipo.int/sme/en/ip_business/collective_marks/collective_marks.htm; and (b) https://www.upcounsel.com/collective-mark
7. (a) https://teamster.org/; (b) https://aflcio.org/; and (c) https://aclu.org
8. (a) https://home.nra.org/; (b) http://www.cotton.org/; (c) https://iupat.org/; (d) http://www.nca-i.com/; and (e) https://www.personalcarecouncil.org/
9. https://euipo.europa.eu/ohimportal/en/certification-marks
10. (a) https://www.ul.com/; (b) https://www.bohanmathers.com/collective-marks-and-certification-marks/; (c) http://www.bis.org.in/other/enforcement.asp; and (d) https://www.iso.org/iso-9001-quality-management.html
11. (a) http://www.sughrue.com/files/Publication/a5e682a6-09e8-4fb4-8d52-f3ba796ee215/Presentation/PublicationAttachment/28d42aa1-f2c4-4516-9a6c-f84323a0b1a7/tradedress.htm; (b) http://corporate.findlaw.com/intellectual-property/trade-dress-the-forgotten-trademark-right.html; (c) https://www.upcounsel.com/trade-dress. See also, References 11(a) and (b); (d) https://tmep.uspto.gov/RDMS/TMEP/Oct2012#/Oct2012/TMEP-1200d1e718.html; and (e) https://www.upcounsel.com/trade-dress
12. (a) See, TMEP 1209.01(a) https://tmep.uspto.gov/RDMS/TMEP/Oct2012#/Oct2012/TMEP-1200d1e7036.html; and (b) A mark whose *intrinsic nature serves to identify a particular source* is considered as an inherently distinctive mark. *Two Pesos v. Taco Cabana*, 505 U.S. 763, 768 (1992).
13. https://www.upcounsel.com/trademark-distinctiveness
14. https://www.drm.com/resources/spectrum-of-distinctiveness
15. https://www.iaca.org/wp-content/uploads/Distinctiveness-Spectrum-Examples-of-Marks.pdf

16. https://www.quora.com/What-is-the-meaning-and-story-behind-the-Starbucks-logo
17. http://thenewswheel.com/behind-badge-mercedes-benz-star-emblem-big-secret/#New-Name--New-Logo
18. (a) See, TMEP 1209.01: https://tmep.uspto.gov/RDMS/TMEP/Oct2012#/Oct2012/TMEP-1200d1e6993.html
19. S. M. McJohn. (2009). *Intellectual Property, Examples and Explanations*. New York: Aspen Publishers, p. 353.
20. (a) See, TMEP 1209.01(b): https://tmep.uspto.gov/RDMS/TMEP/Oct2012#/Oct2012/TMEP-1200d1e6993.html; and (b) See, TMEP 1210.01(a): https://tmep.uspto.gov/RDMS/TMEP/current#/current/TMEP-1200d1e8253.html
21. https://www.iaca.org/wp-content/uploads/Distinctiveness-Spectrum-Examples-of-Marks.pdf
22. See, TMEP 1209.1(c): https://tmep.uspto.gov/RDMS/TMEP/Oct2012#/Oct2012/TMEP-1200d1e6993.html
23. See, page 9, https://www.uspto.gov/sites/default/files/trademarks/law/Trademark_Statutes.pdf
24. See, TMEP 1203.02: https://tmep.uspto.gov/RDMS/TMEP/Oct2012#/Oct2012/TMEP-1200d1e3042.html
25. See, *In re Intex Plastics Corp.*, 215 USPQ 1045, 1048 (TTAB 1982).
26. See, *Stabilisierungsfonds fur Wein v. Peter Meyer Winery GmbH*, 9 USPQ2d 1073, 1076 (TTAB 1988); *In re House of Windsor, Inc.*, 221 USPQ 53, 57 (TTAB 1983), *recon. denied*, 223 USPQ 191 (TTAB 1984).
27. See, *A. F. Gallun & Sons Corp. v. Aristocrat Leather Prods., Inc.*, 135 USPQ 459, 460 (TTAB 1962).
28. See, *Intex Plastics*, 215 USPQ at 1048; *Tanners' Council of Am.*, 204 USPQ at 154–155.
29. See, *The American Heritage® Dictionary of the English Language: Fourth Ed.* 2000.
30. See. TMEP 1203.02(a): https://tmep.uspto.gov/RDMS/TMEP/Oct2012#/Oct2012/TMEP-1200d1e3042.html
31. (a) See, TMEP § 1210.01(c): https://tmep.uspto.gov/RDMS/TMEP/current#/current/TMEP-1200d1e8253.html; and (b) See, page 9, https://www.uspto.gov/sites/default/files/trademarks/law/Trademark_Statutes.pdf
32. (a) See, TMEP §1203.03(c) and TMEP § 1203.03(d): https://tmep.uspto.gov/RDMS/TMEP/Oct2012#/Oct2012/TMEP-1200d1e3042.html; and (b) See, TMEP §1203.03(e) and TMEP §1203.03(f): https://tmep.uspto.gov/RDMS/TMEP/Oct2012#/Oct2012/TMEP-1200d1e3042.html
33. See, TMEP 1203.03(b): https://tmep.uspto.gov/RDMS/TMEP/Oct2012#/Oct2012/TMEP-1200d1e3042.html
34. See, TMEP 1203.03(c): https://tmep.uspto.gov/RDMS/TMEP/Oct2012#/Oct2012/TMEP-1200d1e3042.html
35. See, *In re Lebanese Arak Corp.*, 94 USPQ2d 1215, 217 (TTAB 2010); *In re Squaw Valley Dev. Co.*, 80 USPQ2d 1264, 1267 (TTAB 2006); *Order Sons of Italy in Am. v. The Memphis Mafia, Inc.*, 52 USPQ2d 1364, 1368 (TTAB 1999); *Harjo v. Pro-Football Inc.*, 50 USPQ2d 1705, 1740–1741 (TTAB 1999), *rev'd on other grounds*, 284 F. Supp. 2d 96, 68 USPQ2d 1225 (D.D.C. 2003), *remanded on other grounds*, 415 F.3d 44, 75 USPQ2d 1525 (D.C. Cir. 2005), *and aff'd*, 565 F.3d 880, 90 USPQ2d 1593 (D.C. Cir.2009), *cert. denied*, 130 S. Ct. 631 (2009).
36. https://cyber.harvard.edu/people/tfisher/IP/2010%20Lebanese%20Arak%20Abridged.pdf

37. https://h2o.law.harvard.edu/cases/4830
38. https://www.quimbee.com/cases/greyhound-corp-v-both-worlds-inc
39. See, page 9, https://www.uspto.gov/sites/default/files/trademarks/law/Trademark_Statutes.pdf
40. See, 1202.02(a)(iii)(A); https://tmep.uspto.gov/RDMS/TMEP/Oct2012#/Oct2012/TMEP-1200d1e718.html
41. (a) https://supreme.justia.com/cases/federal/us/532/23/; (b) https://supreme.justia.com/cases/federal/us/514/159/; and (c) https://supreme.justia.com/cases/federal/us/456/844/case.html
42. See, 1202.02(a)(V); https://tmep.uspto.gov/RDMS/TMEP/Oct2012#/Oct2012/TMEP-1200d1e718.html
43. See, TMEP §1209.01(b). https://tmep.uspto.gov/RDMS/TMEP/Oct2012#/Oct2012/TMEP-1200d1e6980.html; See also, Section 5.3.1.2(1) in the above discussion.
44. (a) See, TMEP §§1210.02–1210.02(b)(iv)); (b) See, TMEP §1210.03; (c) See, TMEP §§1210.04–1210.04(d); (d) See, TMEP §1210.04(c); and (e) See, TMEP §§1210.05(c)–(c)(ii); (f) See, http://www.wipo.int/export/sites/www/sct/en/meetings/pdf/wipo_strad_inf_5.pdf
45. http://info.legalzoom.com/trademarks-duration-23713.html
46. http://www.wipo.int/trademarks/en/
47. (a) The *Nairobi Treaty*, which deals with *the obligation to protect the Olympic symbol* (five interlaced rings) – *against use for commercial purposes* (in advertisements, on goods, as a mark, etc.) *without the authorization of the International Olympic Committee* – is not included here; (b) http://www.wipo.int/treaties/en/classification/nice/;(c)http://www.wipo.int/treaties/en/classification/vienna/; (d) http://www.wipo.int/treaties/en/ip/tlt/; and (e) http://www.wipo.int/treaties/en/ip/singapore/
48. See, § 32 (15 U.S.C. § 1114). https://www.uspto.gov/sites/default/files/trademarks/law/Trademark_Statutes.pdf
49. (a) See, Overview: https://www.law.cornell.edu/wex/trademark_infringement; (b) See, breaking down the elements, Reference 49(a); and (c) https://www.uspto.gov/page/about-trademark-infringement
50. https://www.uspto.gov/sites/default/files/trademarks/law/Trademark_Statutes.pdf
51. (a) http://www.worldtrademarkreview.com/Magazine/Issue/40/Country-correspondents/United-States-Edwards-Wildman-Palmer-LLP; and (b) See, Reference 51(a) for the benefits of registration as well.
52. See, TMEP 906.02. https://tmep.uspto.gov/RDMS/TMEP/Oct2012#/result/TMEP-900d1e1351.html?q=Fraud&ccb=on&ncb=off&icb=off&fcb=off&ver=Oct2012&syn=adj&results=compact&sort=relevance&cnt=10&index=3
53. See, TMEP 906.04. https://tmep.uspto.gov/RDMS/TMEP/Oct2012#/Oct2012/TMEP-900d1e1447.html
54. (a) http://smallbusiness.findlaw.com/business-operations/internet-cybersquatting-definition-and-remedies.html; and (b) https://www.techopedia.com/definition/2393/cybersquatting
55. See, the *Anti-Cybersquatting Consumer Protection Act* (ACPA). https://www.gpo.gov/fdsys/pkg/CRPT-106srpt140/html/CRPT-106srpt140.htm
56. ICAN uses the *Uniform Domain Name Dispute Resolution Policy* (UDNDRP), an international policy designed for arbitration of domain name disputes instead of litigation

6

Other Forms of IPRs

I personally think intellectual property is an oxymoron. Physical objects have a completely different natural economy than intellectual goods. It's a tricky thing to try to own something that remains in your possession even after you give it to many others.

John Perry Barlow

This chapter deals with other forms of intellectual property rights (IPRs), not so far covered in chapters 3–5, such as listed as follows:

1. *Industrial designs,*
2. *Plant varieties,*
3. *Geographical indications,*
4. *Semiconductor integrated circuits layout-designs,* and
5. *Traditional knowledge.*

It must be pointed out upfront that the forms of IPRs covered in chapters 3–5—namely, *patents, copyrights,* and *trademarks*—exist in the IP systems of all the countries and regions interested in protecting their intellectual property. However, the above forms of IPRs do not retain their identities in all the countries and may be considered for protection within the categories of *patents, copyrights,* and *trademarks.* For instance, in the United States of (USA), the *industrial designs, plant varieties,* and *semiconductor integrated layout designs* are considered within the category of patents, while *geographical indications* are dealt with under the category of *trademarks.*

Industrial Designs

In the European Union (EU), *industrial designs* are a class of IP.[1a] India also has *industrial designs.*[1b] In the United Kingdom, they are simply known as the *designs.*[1c] In the USA, these are NOT a separate form of IP, but considered within the category of patents, namely—the *design patents.*[1d] China also has *design patents.*[1e] Regardless, all of them refer to similar subject matter of IP, in spite of different names.

Industrial Designs—Definitions

According to World Intellectual Property Organization (WIPO)[2a]:

> An *industrial design* refers to *ornamental or aesthetic aspects of an article.* An industrial design may consist of three-dimensional features, such as the shape of an article or two-dimensional features, such as patterns, lines or color.

This definition is also applicable to the designs in the United Kingdom.

According to the United States Patent and Trademark Office (USPTO)[2b]:

> A *design* consists of the *visual ornamental characteristics embodied in, or applied to, an article of manufacture.* Since a design is manifested in appearance, the subject matter of a design patent application may relate to the configuration or shape of an article, to the surface ornamentation applied to an article, or to the combination of configuration and surface ornamentation. A design for surface ornamentation is inseparable from the article to which it is applied and cannot exist alone. It must be a definite pattern of surface ornamentation, applied to an article of manufacture.

Generally speaking, a "design patent" protects the way an article looks (35 U.S.C. 171), while a "utility patent" protects the way an article is used and works (35 U.S.C. 101).[2b] Both design and utility patents may be obtained if the invention comprises of both aspects: the *utility* and *ornamental appearance.*

According to the Designs Act 2000 of India[2c]:

> A *design* means only *the features of shape, configuration, pattern, ornament, or composition of lines or colors applied to any article* whether in two dimensional or three dimensional or in both forms, by any industrial process or means, whether manual, mechanical or chemical, separate or combined, which in the finished article appeal to and are judged solely by the eye, but does not include any mode or principle of construction or anything which is in substance a mere mechanical device, and does not include any trade mark as defined in clause (v) of sub-section (1) of section 2 of the Trade and Merchandise Marks Act, 1958 or property mark as defined in section 479 of the Indian Penal Code or any artistic work as defined in clause (c) of section 2 of the Copyright Act, 1957.

According to China's Patent Law[2d]:

> A *design* refers to *the shape, pattern, or the combination thereof, or the combination of the color with shape and pattern, which are rich in an aesthetic appeal*

> *and are fit for industrial application.* While most items with a distinctive exterior appearance can be covered by a design patent, the following cannot: 2-dimensional trademarks, parts of a design which cannot be used or sold separately, and items which contravene local law or have a negative effect on public interest.

The Industrial Design Right

As per law, the owner of a *registered industrial design* (in Europe) or of a *design patent* (in the USA) has *the right to prevent third parties from making, selling, or importing articles bearing or embodying a design which is a copy, or substantially a copy, of the protected design, when such acts are undertaken for commercial purposes.*[3]

In general, an industrial design needs to be registered in order to receive protection under the industrial design law. They are referred to as the "registered designs." In the USA, however, the U.S. patent laws protect the "design patents." In some countries, the industrial design laws grant—limited protection to so-called *unregistered industrial designs,* without registration. Further, in some countries, industrial designs may also be protected as works of art under copyright law. In the USA, a design patent protects *only the appearance of the article* and *not structural or utilitarian features.* The principal statutes (United States Code) governing design patents are: 35 U.S.C. §§ 171–173; §§ 102, 103, 112, and 132. See, also elsewhere.[4a]

As a rule, a design must be either *new* or *original* to be registrable. The assessment of *novelty* and *originality* depends on the country's laws. In general, an industrial design is considered *new* or *novel* if it has not previously been disclosed to the public; it may be considered *original* if it significantly differs from known designs or combinations of known design features.

Term of Protection of Industrial Designs

Generally speaking, *industrial design rights* are granted only for a limited period. While the duration of the term of protection of industrial designs varies from one country to another, they *last at least for* 10 years. Further, in several countries, the total term of protection is divided into successive renewable periods. Thus, in Europe, a registered community design is initially valid for 5 years from the date of filing and can be renewed in blocks of 5 years up to a maximum of 25 years.[4b] Similarly, an unregistered community design is given protection for a period of 3 years from the date on which the design was first made available to the public within the territory of the European Union. After 3 years, the protection cannot be extended.[4b] India grants 10 years of protection for industrial designs in the country of filing, with one additional renewal possible for 5 more years.[4c] On the contrary, China's design patents last for a maximum of 10 years and are not renewable.[4d] Finally, in the USA, the design patents are granted for a term of 14 years from the date of grant.[4e]

Products Covered by Industrial Designs

Industrial designs are applicable to an exceptionally wide array of *industrial products* and *handicrafts*: from packages and containers to textile designs, from leisure goods to home appliances, from lighting equipment to jewelry, from technical and medical instruments to watches, from vehicles to architectural structures, and from electronic devices to graphic symbols, graphical user interfaces, and logos, to mention a few.

Benefits of Industrial Designs

An industrial design enables a product to become *more attractive and appealing to customers*. In fact, *the appearance of a product* can be a key factor in the *customer's purchase decision*. In some cases, the *design can be so effective that both appearance and functionality work synergistically* (though functionality is protected by a patent). For example, when Apple iPhone was introduced into the market, its design was so effective that many customers felt that the product was *not only appealing to own, but also easy to operate*. Most importantly, when designed right, *the looks and functionality of the product become inseparable in the minds of consumers*. Thus, when Apple introduced the first generation of iPhones in 2007, it sold almost 1.4 million units worldwide in the first year. However, over the next 8 years, the sales of the Apple iPhone soared, reaching 230 million units by 2015.[5] This is the *design-based innovation approach*.

At one time, design was just an afterthought in the product development process of most companies. However, this situation has evolved over the years, and design evokes entirely new business meaning today. In fact, the following conclusions emerged about "design" from one of the design leaders in the consumer products world[6]:

1. Design *adds value* across the entire spectrum of a product's development—from concept to commercialization;
2. Design helps *in developing packaging, identity assets*, and *artwork*;
3. Design-based research and development at the front end (this can be either *closed* or *open innovation*), helps *to create innovative solutions for complex consumer needs*; and
4. Design helps to improve the form of a product (*how it looks*), the functionality of a product (*how it works*), and its meaning for consumers (*how the brand connects emotionally and engages with the consumer*).

Thus, industrial designs can be very important to both small- and medium-sized enterprises and large companies alike, irrespective of the sector of their activity. Industrial designs are relatively inexpensive to develop and protect. Therefore, an effective protection of industrial designs can *promote innovation* and *contribute to economic growth and development*.

Industrial Design Related Treaties

The international legal framework for industrial designs is constituted by the treaties that WIPO administers, along with the national and regional laws. The treaties are:

1. The Hague Agreement;
2. The Locarno Agreement;
3. Paris Convention; and
4. WIPO Convention.

A brief description of the details of the treaties are as follows:

The Hague Agreement

This agreement *governs the international registration of industrial designs.* The Hague Agreement, first adopted in 1925, effectively establishes an international system—*The Hague System—that allows industrial designs to be protected in multiple countries or regions with minimal formalities.* According to the WIPO[7a]:

> The *Hague Agreement* allows applicants to register an *industrial design* by filing a single application with the *International Bureau* of WIPO, enabling design owners to protect their designs with minimum formalities in multiple countries or regions. The *Hague Agreement* also simplifies the management of an *industrial design* registration since it is possible *to record subsequent changes* and *to renew the international registration through a single procedural step.*

Two Acts, namely—the 1999 Act and the 1960 Act—of The Hague Agreement, are currently in operation. In 2009, the application of the 1934 Act of The Hague Agreement was frozen, *to simplify and streamline the overall administration of the international design registration system.*

While the 1960 Act is open to the nations party to the Paris Convention, *the governments of prospective contracting parties are encouraged to join the more advantageous 1999 Act.*

The Locarno Agreement

The *Locarno Agreement*[7b,d] launched a classification for industrial designs, known as the *Locarno Classification.*[7c] This agreement concluded at Locarno in 1968 and was later amended in 1979. The Locarno Classification comprises:

1. A list of classes and subclasses;
2. An alphabetical list of goods which constitute industrial designs, with an indication of the classes and subclasses into which they fall;

3. Explanatory notes; and
4. The original list of classes and sub-classes that was attached to the Locarno Agreement when it was adopted.

The Locarno Classification is "solely of an administrative character" and is not binding on the contracting countries "in terms of the nature and scope of the protection that designs receive in those countries."[7d]

The Paris Convention

See chapter 2 for full details.

The WIPO Convention

The WIPO Convention is an instrument of the WIPO; it was signed on July 14, 1967 (at Stockholm), it came into force in 1970, and was amended in 1979. The WIPO is an intergovernmental organization that became one of the specialized agencies of the United Nations system of organizations in 1974.[7e] The WIPO has two main goals: (i) to *promote the worldwide protection* of intellectual property and (ii) to *ensure that the intellectual property unions are administratively cooperating with each other,* as per the treaties that the WIPO administers.

Accordingly, the WIPO plays a dual role performing, on one hand, the administrative tasks of the unions, while on the other, leading many activities, including: (i) *the setting up of norms and standards* for the protection and enforcement of IPRs through the conclusion of international treaties; (ii) *conducting program activities,* to offer legal and technical assistance to states in the field of IPRs; (iii) *leading the international classification, standardization, and documentation activities,* with the cooperation of IP offices on patents, trademarks, and industrial designs; and (iv) *ensuring the registration and filing activities, involving services related to international applications for patents and the registration of marks and industrial designs.*

Plant Varieties

In the EU, "plant variety" is a form of IP.[8a] According to the China IPR Helpdesk, "In 1997, China promulgated its first New Varieties of Plants (NVPs) protection law, *The Regulations of the People's Republic of China on the Protection of New Varieties of Plants* (the China NVPs Regulations), and the respective implementation of rules relating to agriculture and forestry. In 1999, China joined the Convention of the International Union for the Protection of New Varieties of Plants (UPOVs). As a result, China protects NVPs from foreign countries who are also members of UPOVs, this includes most European countries."[8b] India also recognizes this form of IP and enacted the *Protection of Plant Varieties and Farmers' Rights Act* in 2001.[8c] In the USA, this form of IP is protected by the *Plant Patents.*[8d]

Plant Varieties—Definitions

According to the UPOVs[9a,b]:

> "Plant Variety Protection, also known as, Plant Breeder's Right" (PBR) *is an IPR granted to the breeder of a new plant variety. PBR is an independent "sui generis" form of protection, tailored to protect new plant varieties and has certain features in common with other intellectual property rights.*

In China[9c]:

> New Variety of Plant (NVP) refers to *improved plant varieties, cultivated or developed from wild plants discovered by breeders, which possess novelty, distinctness, uniformity, and stability, as well as an adequate denomination.* The protection of NVP grants the plant breeders the exclusive rights to use the variety for commercial use, in recognition of the special efforts made by the breeders in developing new varieties for agriculture, horticulture, and forestry.

India already initiated the procedure for acceding to the UPOV Convention in 2017.[9a] Currently, the *Protection of Plant Varieties and Farmers' Authority* of India recognizes and protects the rights of the farmers in respect of their contribution made at any time in conserving, improving, and making available plant genetic resources for the development of the new plant varieties.[9d]

In America, the USPTO grants a *plant patent* to an inventor (or the inventor's heirs or assigns) *who has invented or discovered and asexually reproduced a distinct and new variety of plant, other than a tuber propagated plant or a plant found in an uncultivated state.* This grant protects the *patent owner's right* to exclude others from asexually reproducing the plant, and from using, offering for sale, or selling the plant so reproduced, or any of its parts, throughout the United States, or from importing the plant so reproduced, or any part thereof, into the United States.[10]

Protection of New Varieties of Plants

Several countries, including developing countries and countries in transition to a market economy, have already been either making serious efforts to introduce the *Protection of New Varieties of Plants* system or have already introduced it, based on the International Convention for the *Protection of New Varieties of Plants* (UPOV Convention) *in order to provide an effective, internationally recognized system.* With regards to the Protection of New Varieties of Plants system, UPOV has clarified that its mission is: "To provide and promote an effective system of plant variety protection, with the aim of encouraging the development of new varieties of plants, for the benefit of society."[11a]

Under the UPOV Convention: "the breeder's right is only granted where the variety is *new, distinct, uniform, stable,* and has a *suitable denomination.* The breeder's right does not extend to the acts done: (i) privately and for non-commercial purposes; (ii) for experimental purposes; and (iii) for the purpose of breeding other varieties."[11b]

The creation of the UPOV *System of Plant Variety Protection* and introduction of the UPOV *membership,* are intended *to effectively capitalize on a variety of innovation opportunities*[12]:

1. Increased breeding activities;
2. Greater availability of improved varieties;
3. Increased number of new varieties;
4. Diversification of types of breeders (e.g., private breeders, researchers);
5. Increased number of foreign new varieties;
6. Encouraging the development of a new industry competitiveness on foreign markets; and
7. Improved access to foreign plant varieties and enhanced domestic breeding programs.

In the USA, as this IP is protected under the *plant patents,* the *provisions and limitations* are very clearly laid down. Thus, according to 35 U.S.C. § 161[13]:

> Whoever *invents* or *discovers* and *asexually reproduces any distinct and new variety of plant,* including cultivated sports, mutants, hybrids, and newly found seedlings, *other than a tuber propagated plant or a plant found in an uncultivated state,* may obtain a patent therefore, subject to the conditions and requirements of this title.
>
> The provisions of this title relating to *patents for inventions* shall apply to *patents for plants,* except as otherwise provided.

Further, according to 35 U.S.C. § 161, a plant patent *must also satisfy* the general requirements of patentability, as outlined below[13]:

1. That the plant was invented or discovered in a cultivated state, and *asexually reproduced;*
2. That the plant is not a plant which is excluded by statute, *where the part of the plant used for asexual reproduction is not a tuber food part, as with potato or Jerusalem artichoke;*
3. That the inventor named for a plant patent application *must be the person who actually invented the claimed plant,* for example, discovered or developed and identified or isolated, and asexually reproduced the plant;

4. That the plant has *not been patented, in public use, on sale, or otherwise available to the public* prior to the effective filing date of the patent application with certain exceptions;
5. That the plant has *not been described in a U.S. patent or published* patent application with certain exceptions;
6. That the plant *be shown to differ from known, related plants by at least one distinguishing characteristic,* which is more than a difference caused by growing conditions or fertility levels, etc.; and
7. That the invention *would not have been obvious to one having ordinary skill in the art* as of the effective filing date of the claimed plant invention.

In the USA, a plant patent lasts for 20 years from the date of filing the application.

Also, the plant patent owner's right allows him/her *to exclude others from asexually reproducing the plant, and from using, offering for sale, or selling the plant so reproduced, or any of its parts, throughout the United States, or from importing the plant so reproduced, or any part thereof, into the United States.*

Geographical Indications

In the EU, "geographical indications" (GIs) are a form of IP. Like the EU, India also recognizes GIs as a separate form of IP. In the USA, the USPTO does not have a special register for GIs. However, the USPTO has protected GIs through its *trademark system* for decades, well before the term "geographical indications" came into existence, with the World Trade Organization Agreement on *Trade-Related Aspects of Intellectual Property Rights* (TRIPS). China also follows the USA and protects GIs through its *trademark system.*

In todays' world, consumers are paying increasing attention to *the geographical origin of products* and weighing its relation to quality, reliability, make, and reputation of the products in their purchasing decisions. GIs convey to consumers about *the origin-bound characteristics of a product.* Consequently, GIs have become key to developing brands *that promise quality-bound-to-the-origin of products.* Clearly, GIs serve as *product differentiators* in the market place, to enable consumers to *seek specific products*—whose make, quality, and reliability can be trusted *based on the legend and reputation of the geographical origin.*

"Brand recognition" is key to business success. Some consumers believe that the quality and reliability of a product is rooted in the history and reputation of the "place of origin," and are willing to pay higher prices for such products. Indeed, this has paved the way for the innovation and commercialization of branded, *GI-specific products, which proclaim that the quality and reliability are intimately linked to the place of origin.*

Geographical Indications—Definitions

According to the WIPO[14a]:

> A *Geographical Indication* (GI) is a sign used on products that have *a specific geographical origin* and *possess qualities or a reputation that are due to that origin*. In order to function as a GI, a sign must *identify a product as originating in a given place*. In addition, *the qualities, characteristics, or reputation of the product should be essentially due to the place of origin*. Since the qualities depend on the geographical place of production, *there is a clear link between the product and its original place of production*.

According to the Geographical Indications Registry of India[14b]:

> *Geographical Indications* (GIs) of goods are defined as that aspect of industrial property which refer to the geographical indication *referring to a country* or *to a place situated therein* as being the country or *place of origin of that product*. Typically, such *a name conveys an assurance of quality and distinctiveness* which is essentially *attributable to the fact of its origin in that defined geographical locality, region, or country*.

According to Article 22.1 of the TRIPS Agreement, GIs are[14c]:

> *Indications* which identify a good *as originating in the territory* of a member, or a *region* or *locality* in that territory, *where a given quality, reputation, or other characteristic of the good is essentially attributable to its geographical origin*.

Types of Products Covered by GIs

GIs are typically used for *agricultural products, foodstuffs, wine* and *spirit drinks, handicrafts,* and *industrial products*.[15] The strength of a GI depends on *national policy* and *consumer perception*.

The different categories of GIs and examples for them are as follows:

Agricultural Products

Darjeeling tea, originating in a place known as *Darjeeling* in India, is one of world's most fabled black teas grown at elevations of 600–2000 meters above the sea level. The exceptional quality and reputation of the black tea is traced back to the local climate, soil conditions, altitude, traditional knowledge, and expertise in meticulous processing.[16a]

Foodstuffs

Roquefort cheese, a characteristic blue cheese, is made in a region in southwest France, around the municipality of *Roquefort-sur-Soulzon*. This cheese is smooth and compact, with even blue veins, a very distinctive aroma, slight scent of mold, and a fine robust taste. It is made from raw, whole sheep's milk from the Lacaune breed. The unique features and taste of *Roquefort cheese* owe their origins to the characteristics of the milk obtained from indigenous breeds of sheep fed according to tradition, the characteristics of the caves in which the cheese is aged, and the traditional know-how used in each step of the cheese making process.[16b]

Wine

Champagne is a unique type of sparkling wine and an alcoholic drink, produced from the grapes grown in the *Champagne region* of France. The special taste of *champagne* owes its origins to a variety of factors including—sourcing of grapes exclusively from the specific parcels in the *Champagne* appellation, strict vineyard practices, specific pressing regimes unique to the region, and methods such as secondary fermentation in the bottle to create carbonation.[16c]

Spirit Drinks

Tequila is a regional *distilled beverage* and type of *alcoholic drink* made from the blue agave plant, primarily in the area surrounding the *city of Tequila*, 65 km (40 mi) northwest of Guadalajara, in the highlands (*Los Altos*) of state of Jalisco, in Mexico. The characteristic taste of tequila owes its origins to the red volcanic soil in the region around the city of Tequila, which is particularly well suited for growing the blue agave, which are large in size and sweet in aroma and taste.[16d]

Handicrafts

Kondapalli bommalu (*meaning*: Kondapalli toys) are *historically traditional handicrafts made from a special soft wood* in the *Kondapalli village of Krishna District*, Andhra Pradesh, India. *Bommala colony* (*meaning*: toys colony) in Kondapalli Village is where the art of crafting takes place. The Kondapalli *Bommalu* are made from *Tella Poniki*—a uniquely soft wood which is extremely light in weight—found in nearby *Kondapalli Hills*. The wood is first carved out, its edges smoothed, and finally colored with oil and watercolors or vegetable dyes and enamel paints, to create the *Kondapalli Bommalu*.[16e]

Industrial Products

Swiss made watches convey that the watches are *manufactured in Switzerland according to the tradition, know-how, and quality criteria of Swiss watchmaking*, which enjoys a great reputation around the world. According to the Federation of the Swiss Watch Industry, the goal is to "guarantee satisfaction of the consumer who, when buying a Swiss-made watch, expect it to correspond to the quality and the reputation of Swiss watchmaking tradition and therefore to be manufactured in Switzerland and to incorporate a high added value of Swiss origin."[16f]

Similarities and Differences between GIs and Trademarks[17]

Similarities

1. ***Distinctiveness***. Both GIs and trademarks can be *distinctive*; and
2. ***Primary purpose***. Both are used to *distinguish* goods/services in the marketplace. Both are intended to *identify the source of the good/service* and *setup customer expectations of quality and reliability* of the good/service.

Differences

1. ***Source***. Trademarks inform customers of the *source company originating* the good/service. Further, they enable the customers to build associations between *the good/service, its quality,* and *the company*. GIs, on the other hand, inform customers about the *place of origin* (and more importantly, its tradition and background) for the *good*. GIs let the customers build associations between a good, *its unique originating place and traditions,* and *assurance of quality, reliability, and distinctiveness*.
2. ***Owners v. users***. A trademark is a fanciful, arbitrary, or suggestive mark that *may be used either by its owner* or *a person who has been authorized*. In contrast, a GI—which corresponds to *the place of origin of the good,* or to *the name by which the good is known in that place*—could be used *by several people in the originating place,* who may be involved in production of the good, according to the collectively agreed standards.
3. ***Assign or license***. A trademark *can be assigned or licensed to anyone, anywhere in the world,* because it is linked to a specific company and not to any particular place. On the contrary, a GI *can neither be assigned nor licensed to anyone outside that place of origin* or *group of authorized producers*.

GI Rights of Authorized Users

A GI right empowers the holder of the right to prevent its use/misuse by a third party whose products do not conform to the specific GI standards. For example, in the jurisdictions where the GI of *Kondapalli Bommalu* is protected, the artists, craftsmen, and businessmen can prevent the use of the term *Kondapalli* for toys not crafted in the *Kondapalli region of India* or not produced as per the standards set forth in the code of practice for the GI.[18]

GI protection is normally obtained *by acquiring a right over the sign that constitutes the indication.* Nevertheless, a protected GI *does not enable the holder of the right to prevent someone else* from making the same product, by using the same exact techniques as those set out in the standards for that indication.

GIs in Different Jurisdictions

GIs are protected in diverse jurisdictions in different ways, often through a combination of two or more approaches (listed below), developed in accordance with the legal, regional, historical, and economic conditions. Thus, the strength of a GI depends on *national policy* and *consumer perception.* GIs are *protected under three different systems*[19a]:

1. Sui generis systems (namely, special regimes of protection);
2. The trademark system, (namely, using *collective marks or certification marks*); and
3. Laws focused on *business practices,* including *streamlined administrative approval schemes for products.*

Some details of the three different systems for GI protection are as follows:

1. ***Sui generis* system.**[19b] A *sui generis* system establishes a *specific* and *exclusive right,* referred to as the *sui generis right* over GIs, that is different from a *trademark right* or *any other* IP right. In the EU, the GIs for wines, spirits, agricultural products, and foodstuffs, are protected by the *sui generis* system. In addition, many jurisdictions—such as India, Switzerland, the countries of Andean Community, and the African Intellectual Property Organization—also use the *sui generis systems* for the GI protection. Unfortunately, however, the terminology used,[19c] in reference to the *sui generis rights* over GIs, is not uniform throughout the world. Thus, different legislations use different terms, such as—*appellations of origin,*[1] *controlled appellations of origin, protected designations of origin, protected geographical indications,* or simply *geographical indications.*[20]

[1] An *appellation of origin* is a special form of GI. Both inform consumers about a product's *geographical origin* and *a quality or characteristic of the product linked to its place of origin.* However, the fundamental difference between a *GI* and an *appellation of origin* is that the link with the place of origin *must be, stronger* in the case of an *appellation of origin,* compared to a GI. The *quality* or *the characteristics* of a product protected as an *appellation of origin* must *exclusively* or *essentially* result from its geographical origin. In other words, in the case of an *appellation of origin, the sourcing of raw materials as well as the processing of the product should happen within the place of origin.* In the case of GIs, any *single criterion attributable to the geographical origin will be sufficient*—whether it is quality, a particular characteristic of the product, or even its reputation.

Generally speaking, an application for registration of a *sui generis right* requires: (a) a *definition of the geographical area* in which the product identified by the GI is produced; (b) a *description of the product's characteristics, quality, or reputation*; and (c) the *standards of production* that the users of the right abide by. Further, certain jurisdictions may require the link to the geographical area be substantiated. Usually, all this information is contained in "the product specification" document. In addition, *sui generis* systems of protection usually require *verification and control schemes* be installed to ensure that users of the GI comply with the agreed standards of production. At a minimum, *sui generis right* protects the holder of GI against any use of the GI by others *to mislead consumers regarding the true geographical origin of the product* or *create unfair competition*. Additionally, *sui generis rights* of wines and spirits afford protection against *any use by an ineligible or unauthorized person*, even when the above criteria are not met. In certain jurisdictions, *sui generis rights* even identify other types of products as well. Finally, some *sui generis systems* protect GIs against *use in a translation, imitation*, or *evocation*.

2. ***Trademark* system**. This system uses *collective marks* or *certification marks* to protect GIs. Some of the countries that use the *trademark system* for protecting GIs are the USA, Canada, China, and Australia. Protection of GIs registered as *collective* or *certification marks* is afforded by the trademark law of the country. The definitions of a *collective mark* and *certification mark* (or, in some countries, guarantee mark) vary from country to country. However, what is universally agreed is that this type of marks: (a) *may be used by more than one person, as long as the users comply with the regulations of use or standards established by the holder* and (b) those regulations or standards may require that the mark *be used only in the case of goods that originate from a particular geographical origin and/or possess specific characteristics.*

   In some jurisdictions, a fundamental difference between *collective marks* and *certification marks* is that the *former may only be used by the members of an association*, while the latter *may be used by anyone who complies with the standards defined by the holder of the mark*. The mark holder—which can be a private or a public entity—*verifies* that the mark is used according to established standards and *certifies* it. Generally, the holder of a *certification mark will not be able to use the mark for one's own purpose.*

3. ***Laws which are focused onbusiness practices***. Under this system, GIs are indirectly protected by laws that specifically focus on business practices, such as laws relating to the *repression of unfair competition, consumer protection*, or *labelling of products*.

Term of Protection for GIs

In *sui generis system*, there is no specific period of validity for registered GIs. Hence, the protection for a registered GI often remains intact, *unless the registration has been cancelled.* GIs registered in the *trademark system* are generally protected in renewable *10-year terms.*

GI Protection Abroad

IPRs are territorial rights. Thus, a GI registered in one jurisdiction may be protected there, but abroad, it runs into the risks of unprotected IP. Consequently, knowing the routes available for the protection of GI abroad is critical. There are four different routes to protection of GIs abroad:

1. Bilateral agreements;
2. The *Lisbon Agreement;*
3. The *Madrid System;* and
4. Direct protection.

Briefly, the details of these options are as follows:

1. ***Bilateral agreements.*** Through *a bilateral agreement,* two states or trading partners (usually customs territories) *may agree to protect each other's* GIs. Such an agreement can be either independent treaties or a part of a wider trade agreement between the parties involved. This is commonly done in the wine and spirits industry.
2. ***The Lisbon Agreement for the protection of appellations of origin and their international registration.***[21] The goal of the *Lisbon Agreement* was to enable the protection of *appellations of origin* at the international level. This is a simple method for obtaining protection for an *appellation of origin* originating in one member state (of the Lisbon Agreement) to the territories of all member states, through "a single international registration." The "country of origin" is "the country/the region or locality situated in the country, whose name constitutes the *appellation of origin* that affords the product its reputation." An internationally registered *appellation of origin,* is protected indefinitely, without renewal. An *appellation* that has been granted protection by a member state cannot be deemed to have become generic in that state as long as it is protected as an *appellation of origin* in the country of origin.
3. ***The Madrid System for the international registration of marks*** (in which the GI is protected in the *country of origin* as a *collective/certification mark*).[22] GIs are protected in several countries as *collective/certification marks* through the *Madrid Agreement concerning the*

international registration of marks and the *protocol relating to the Madrid Agreement*. Both treaties constitute the *Madrid System* and are administered by the WIPO. The *Madrid System* offers the possibility to protect a mark, including a *collective mark* or a *certification mark* or *guarantee mark*, in many countries, by filing one application (an international application) and receiving one registration (international registration). A mark can be the subject of an international application only: (a) if it had already been registered or (b) when the application for registration has been made with the *trademark* office of the contracting party, with which the applicant has the necessary "connection." An international registration is initially valid for 10 years and can be renewed indefinitely in 10-year terms.

4. ***Direct protection***. A GI can also be *directly protected* in the jurisdiction of interest. As mentioned above, GIs are protected by diverse methods in different jurisdictions, sometimes, by a combination of two or more methods. The GI owners may protect their GIs in the target jurisdiction by: (a) either using the method that is available or (b) employing the method that is best suited to their needs where more than one mode of protection is available. For example, to protect a GI in Australia, China, or the United States of America, an application for registration of a *collective/certification mark may be filed directly with the respective trademark offices*. On the other hand, when GI protection is sought in the EU for GIs identifying agricultural products or foodstuffs, one can apply for registration of a *protected geographical indication* or a *protected designation of origin*. Finally, it is also possible to file an application for registration of a *collective mark* with the European Union Intellectual Property Office.

Conditions for GI Protection

In order for a *sign* to qualify as a GI *there should not be any obstacles for registering a* GI *under the applicable law*. Broadly speaking, a necessary condition for this is the fact that *the good identified by the GI needs to have a link to the geographical origin*. This requirement will be met when one can clearly establish the link between the good (identified by the GI) and a *certain quality, reputation, or other characteristic of the geographical origin*. In many jurisdictions, *a single criterion attributable to the geographical origin is sufficient*, whether it is quality, reputation, or a specific characteristic of the product.

Benefits of GI Registration

Registration of a GI protects it and enables the holders of GI (namely, those *who have the right to use the indication*) to *take action against infringers* (namely, those *who use other's GI without authorization*, seeking to *benefit from its reputation*).

A GI possessing a great reputation is a valuable, collective and intangible asset. Although one can possibly use an indication without protection, in an unrestricted manner, its value is easily diminished and eventually lost, without proper protection. In fact, the transformation of a GI into a *generic mark* may occur in different countries and at different times. Therefore, protecting a GI through the WIPO's *Lisbon System* (see, Section F) can prevent registration of the indication as a *trademark* by a third party and also limit the risk of degradation of the GI into a generic term.

In general, GIs that are rooted in a sound business strategy, can:

1. Strengthen *the brand internationally;*
2. Enhance *product differentiation;*
3. Improve *export opportunities for the product;* and
4. Create *sustainable competitive advantage.*

Integrated Circuits Layout Designs

The *integrated circuit layout designs* (ICLDs) are a form of IP. In the USA, an *integrated circuit's* ("chip") (2- or 3-dimensional) *layout design/topography* is known as—"mask work."[2] Thus, *mask work* (ICLD) is—*the layout or arrangement of semiconductor devices such as transistors, resistors, and interconnections* on *a chip.*

Interestingly, ICLDs cannot be effectively protected under *copyright law* or *patent law* for different reasons. Thus, ICLDs cannot be effectively protected under the *copyright law* (except perhaps as decorative art) because a *mask work's geometry is functional nature.*[23a] Similarly, ICLDs cannot also be effectively protected under the *patent law* because the individual *lithographic mask works are not clearly patentable subject matter,*[23b] although any methods/processes associated with the works are patentable. Therefore, since the 1990s, many national governments have been granting "copyright-like exclusive rights" for ICLDs *that confer exclusivity to reproduction of a particular layout, for a limited time.*

Therefore, an international diplomatic conference was held at Washington, DC, in 1989, which adopted a *Treaty on Intellectual Property in Respect of Integrated Circuits,* also called *the Washington Treaty or IPIC Treaty.*[23c] The *Washington Treaty,* signed on May 26, 1989, is open to all the member states of

[2] In photolithographic processes, a mask, or more precisely, a photomask, allows/blocks the light at specific locations in order to create multiple etched layers within actual ICs, sometimes simultaneously, for a multitude of chips on a wafer. For legal definitions of a "mask work," see "Definitions" below.

the WIPO, or the United Nations, and the intergovernmental organizations that meet the required criteria.[23c–e]

In 1995, the *Washington Treaty* has been incorporated by reference into the *TRIPS Agreement of the World Trade Organization.*[23d]

Integrated Circuits Layout Designs—Definitions

According to the *Washington Treaty,* Article 2[23c,d]:

> An *integrated circuit* means *a product,* in its final form or an intermediate form, *in which the elements, at least one of which is an active element, and some or all of the interconnections are integrally formed in and/or on a piece of material* and which is intended to perform an electronic function and,
>
> A *layout-design* (or topography) means "*the three-dimensional disposition,* however expressed, *of the elements, at least one of which is an active element, and of some or all of the interconnections of an integrated circuit,* or *such a three-dimensional disposition prepared for an integrated circuit* intended for manufacture."

According to 37 CFR § 150.1(d)[24a]:

> A *mask work* means *a series of related images,* however *fixed* or *encoded*:
>
> A. Having or representing the predetermined, *three-dimensional pattern of metallic, insulating, or semiconductor material present or removed from the layers of a semiconductor chip product*; and
> B. In which series the relation of the images to one another is that *each image has the pattern of the surface of one form of the semiconductor chip product.*

In India, according to the *Semiconductor Integrated Circuits Layout-Design* (SICLD) *Rules 2000*[24b]:

> A *semiconductor layout design* means a *layout of transistors and other circuitry elements and includes lead wires connecting such elements and expressed in any manner in semiconductor integrated circuits.*

In October 2001, the State Council of the People's Republic of China enacted Regulations on the Protection of Integrated Circuits Layout-Designs.[24c] China adopted the definitions and language of the *Washington Treaty* in the regulations on ICLDs, as it is a member of the World Trade Organization and signatory to the *TRIPS Agreement.* India has been actively trying to protect ICLDs since 2000. Hence, it may be useful to study India's ICLD system.[25]

Criteria for Registration under SICLD Act 2000

According to this act, any ICLD that meets the following criteria can be registered:

1. Original;
2. Distinctive;
3. Capable of distinguishing from any other layout design; and
4. Have not been commercially exploited *anywhere in India* or *in a convention country.*

The *creator* of a layout design *can apply for registration* in India, *if the individual* is:

- An Indian national;
- The national of another country that accords a similar right to the citizens of India; or
- The national of another country *having a principal place of business in India,* or *has place of service in India, in case he/she does not carry out business in India.*

Procedure for Registration

An application for the registration of a layout design shall be made to the registrar on *form LD-1* accompanied with *registration fees* (5000 Indian Rupees) and *three set of drawings* or *photograph of mask layout,* which describe the layout design applied for registration. The steps involved in the ICLD registration process are:

1. Application filing by the *creator of the layout-design* to the SICLD Registry;
2. Application acceptance;
    a. Registrar may *accept/refuse* the application; or *accept it with modifications;* and
    b. The *accepted applications* shall be advertised within 14 days of acceptance.
3. Any *notice of opposition* can be filed within 3 months *from the date of advertisement;*
4. The *counter-statement* to the *notice of opposition* should be filed within 2 months from the date of receipt of copy of *notice of opposition* from the registrar;
5. A copy of the *counter statement* provided to the opposing party;

6. The registrar may hear the parties involved;
7. The registrar will then decide on *the originality of the layout-design* and *grant/reject the application for registration* based on own conclusions; and
8. The *aggrieved party* can appeal to appellate board, or in its absence, civil court for relief on any ruling of the registrar.

Term of Validity of Registration of an ICLD

The term of validity of a registration of ICLD is 10 years:

- From *the date of filing an* ICLD *application for registration*; or
- From *the date of first commercial exploitation* of ICLD in India, or *in any convention country*, or *any country specified by the Government of India*, whichever is earlier.

Key Sources of Information

The fact that effective protection of this emerging form of IP, the ICLDs, is challenging under the *copyright law*, as well as *patent law*, requires one to know the key sources of information to understand its full complexities. In this regard, some authoritative sources of information are: (i) 17 U.S.C. §§ 901–914[26a] and (ii) 37 C.F.R. Sub Chapter C, §§ 150.1–150.6[26b] and *The 1989 Washington Treaty on Intellectual Property in Respect of Integrated Circuits.*[26c]

Traditional Knowledge

> The Congress shall have the power to promote the progress of science and useful arts by securing for limited times to authors and inventors the exclusive right to their respective writings and discoveries.
>
> **Article 1, section 8 of the United States Constitution**

The current system of IPRs evolved from its early medieval beginnings in order to safeguard creativity and innovation; indeed, this legal foundation was critical to catalyze and sustain the growth of the Industrial Revolution. Thus, when *printing* was invented, published books and other written materials became part of the public domain. However, in the fifteenth century, the Republic of Venice initiated the grant of book privileges to the authors of works, which soon became an *accepted norm* in a number of other European countries. Similarly, the first U.S. Patent, signed by President George Washington, was issued to Samuel Hopkins on July 31, 1790 for a process of making potash, an ingredient used in fertilizer. Since then, more than 6 million patents have been issued by the USPTO. Thus, the current IPR system has evolved over the past several hundred years to effectively protect the inventions and artistic works in the industrial west. Over the past few decades, however, *the twin forces of knowledge revolution and globalization* increased IPR awareness *among the indigenous people, local communities, and governments* of emerging nations such as India and China to demand protection for their *Traditional Knowledge Systems.*[27a]

"Traditional knowledge" (TK) does not derive its name based on the length of its existence. In fact, "traditional knowledge" signifies:—(a) that it is *a living body of knowledge,* (b) that it is *valued, contemporized, and carried forward seamlessly from one generation to another,* and (c) that it often *shapes cultural and/or spiritual identity of the people.* Unfortunately, the present system of IPRs is not equipped with effective legal instruments that can protect this kind of TK.

Therefore, it would set a new milestone in international IP law when the *traditional outputs of creativity and innovation* are also recognized as *protectable intellectual property,* enabling key stakeholders—such as, the indigenous people, local communities, and national governments—*to have control over and benefit from their* TK. Indeed, effective protection of TK would also *discourage the misappropriation of an important living body of knowledge,* such as traditional remedies, indigenous art, and music, and *enable the stakeholders to gain control and collectively benefit from their commercial exploitation.*[27a]

Consequently, in Fiscal Year (FY) 2000, the WIPO established an intergovernmental committee (IGC) on related groups of IP, namely—TK, "genetic resources" (GR), and "traditional cultural expressions" (TCE). Nine years later, IGC embarked on a mission to develop *effective international legal instruments* (or *instrument) for protecting* TK, GR, and TCE.

There are three concepts that have appeared in the above discussion that must be described first and defined later. They are:

1. TK;
2. GR; and
3. TCE.

Traditional Knowledge—Definitions

Traditional Knowledge (TK)

The WIPO describes the term,[28a]

> "Traditional knowledge," by referring to a broad subject matter, that includes:
>
> 1. *Tradition-based* literary/artistic/scientific works; performances; inventions; scientific discoveries; designs; marks/names/symbols; undisclosed information; and

2. *All other tradition-based* innovations and creations resulting from intellectual activity in the industrial, scientific, literary, or artistic fields.

"Tradition-based" describes "knowledge systems, creations, innovations, and cultural expressions (TCE), *which have generally been transmitted from generation to generation;* are generally regarded *as pertaining to a particular people* or *its territory;* and are *constantly evolving* in response to a changing environment."

The *subject matter categories* of TK include:

1. Agricultural knowledge; scientific knowledge; technical knowledge; ecological knowledge; medicinal knowledge, including related medicines and remedies; and biodiversity-related knowledge;
2. "Expressions of folklore" (TCE) in the form of music, dance, song, handicrafts, designs, stories, and artwork; elements of languages, such as names, geographical indications, and symbols; and, movable cultural properties; and
3. Excluded from this description of TK would be *items not resulting from intellectual activity in the industrial, scientific, literary or artistic fields,* such as human remains, languages in general, and other similar elements of "heritage" in the broad sense.

Given the complex and dynamic nature of TK, the WIPO argues that *it may not be possible to develop a singular and exclusive definition of the term.*

Consequently, according to the WIPO[28b]:

There is as yet no accepted *definition* of "Traditional Knowledge" (TK) at the international level. Thus, "Traditional Knowledge" (TK):

- Generally, includes—as a broad description of subject matter—*the intellectual and intangible cultural heritage, practices and knowledge systems of traditional communities, including indigenous and local communities;*
- Embraces—in a general sense—*the content of knowledge itself* as well as *traditional cultural expressions* (TCE), *including distinctive signs and symbols associated with traditional knowledge;*
- Refers to—in the narrow sense—*knowledge in a traditional context, as such,* in particular *the knowledge resulting from intellectual activity, and includes know-how, practices, skills, and innovations;* and
- Can be found in a wide variety of contexts, including—*agricultural knowledge; scientific knowledge; technical knowledge; ecological knowledge; medicinal knowledge, including related medicines and remedies; and biodiversity-related knowledge, etc.* (GR).

Further, the Article 1 on "Subject Matter of Protection," of the "Draft Articles on *Traditional Knowledge* Prepared by IGC (July 18–22, 2011)," as incorporated in the document, "Matters Concerning the IGC on *Intellectual Property and Genetic Resources, Traditional Knowledge and Folklore* (WO/GA/40/7)," provides the following definitions:

Definition 1[28c]: "Traditional Knowledge" (TK) refers to *the know-how, skills, innovations, practices, teachings, and learning, resulting from intellectual activity and developed within a traditional context*; and

Definition 2[28c]: "Traditional Knowledge" (TK) is the *knowledge that is dynamic and evolving, resulting from intellectual activities which is passed on from generation to generation* and includes but is not limited to *the know-how, skills, innovations, practices, processes, and learning and teaching, that subsist in codified, oral, or other forms of knowledge systems*. "Traditional Knowledge" (TK) also includes *knowledge that is associated with biodiversity, traditional lifestyles, and natural resources.*

Genetic Resources (GR)

As in the case of TK, the concept of GR[29a] must be explained before its definitions can be noted. GR are defined in the *Convention on Biological Diversity* (CBD). They are *parts of biological materials* that:

1. Contain genetic information of value; and
2. Are capable of reproducing or being reproduced.

Examples include—*materials of a plant origin, animal origin, or microbial origin,* such as—medicinal plants, agricultural crops, and animal breeds.

It is important to know that *some* TK *is closely associated with* GR through:

1. The *utilization* and *conservation* of the resource, often *over generations;* and
2. Their *common use* in modern scientific research,

because TK often provides leads to isolate valuable active organic compounds within GRs.

Therefore, in and of themselves, GR *are not regarded as intellectual property, as they are not creations of the human mind.* However, *inventions* either *based on* GR (whether or not they are linked to TK) or *developed using* GR *can be IP, namely, patentable* or *protectable by plant breeders' rights.* In this regard, the WIPO's work on GR is complementary to the international legal framework—that regulates the access to, and benefits from the utilization of GR—which is defined by the CBD, and its *Nagoya Protocol,* and the *International Treaty on Plant Genetic Resources for Food and Agriculture* of the UN *Food and Agriculture Organization* (FAO). The definitions of GR are as follows[29a]:

Definition 1: Article 2 of the CBD (1992)[29b] defines GR as *"genetic resources, organisms, or parts thereof, populations, or any other biotic component of ecosystems with actual or potential use or value for humanity;"*

Definition 2: Article 1 of the Decision 391 on *Common Regime on Access to Genetic Resources of Andean Community* (1996)[29c] defines GR broadly as *"all biological material that contains genetic information of value or of real or potential value;"* and

Definition 3: Article 2 of the FAO *International Treaty on Plant Genetic Resources for Food and Agriculture* (2001)[29d] defines "Plant Genetic Resources" as *"any material of plant origin, including reproductive and vegetative propagating material, containing functional units of heredity."*

Traditional Cultural Expressions (TCE)

The WIPO uses two separate terms, namely—TCE,[30a] and "Expressions of Folklore" (EF)—to refer to the *tangible* and *intangible forms* in which "Traditional Knowledge" and culture are *expressed, communicated,* or *manifested*. Frequently, these two terms TCE and EF are used interchangeably, and therefore, the use of simply TCE for both may be prudent. Examples for TCE include traditional music, performances, narratives, names and symbols, designs, and architectural forms.

It is important to note here that *the use of these terms in no way suggests any consensus* among the WIPO Member States *on the validity or appropriateness of these or other terms*. Also, it does *not affect or limit the use of other terms* in national or regional laws. The WIPO IGC uses the following descriptions of TCE in all its current deliberations[30b]:

Description 1

1. Traditional Cultural Expressions (TCE) *are any form of artistic expression, tangible or intangible,* in which traditional culture (and knowledge) are embodied including, but not limited to: (a) phonetic or verbal expressions; (b) musical or sound expressions; (c) expressions by action; and (d) tangible expressions of art;
2. Protection extends to Traditional Cultural Expressions (TCE) that are: (a) *the result of creative intellectual activity*; (b) *passed from generation to generation*; (c) distinctive of or the unique product of the cultural and social identity and cultural heritage; and (d) maintained, used, or developed by the beneficiaries as set out in Article 2.[30b]
3. The terminology used to describe the protected subject matter should be determined at the national, regional, and sub-regional levels.

Description 2

1. Traditional Cultural Expressions (TCE) are *any form of expressions, tangible or intangible, or a combination thereof, which are indicative of traditional culture and knowledge and have been passed on from generation to generation, including, but not limited to*:
   a. Phonetic or verbal expressions, such as stories, epics, legends, poetry, riddles, and other narratives; words, signs, names, and symbols;
   b. Musical or sound expressions, such as songs, rhythms, and instrumental music, the sounds which are the expression of rituals;
   c. Expressions by action, such as dances, plays, ceremonies, rituals, rituals in sacred places and peregrinations, traditional sports and games, puppet performances, and other performances whether fixed or unfixed; and
   d. Tangible expressions, such as material expressions of art, handicrafts, works of mas, architecture, and tangible spiritual forms, and sacred places.
2. Protection shall extend to any traditional cultural expression that is associated *with the cultural and social identity of the beneficiaries* as defined in Article 2, and is used, maintained, or developed by them as part of their *cultural or social identity or heritage* in accordance with national law and customary practices;[30b] and
3. The specific choice of terms to denote the protected subject matter should be determined by national legislation.

Indigenous Knowledge and Heritage

Thus, there are two more important concepts, namely—"indigenous knowledge" (IK)[31] and "heritage" (H)[31]—that must be explained, before we can arrive at a *holistic picture* provided by the WIPO's research on TK, TCE, GR, IK, and H.

Indigenous Knowledge (IK)

The question: "Who are the indigenous people?" attracts a considerable attention, discussion, and debate. IK[31] may be described as *the knowledge held and used by communities and peoples that are "indigenous" to a place, region, or country*. The description of the term "indigenous," in *the Study of the Problem of Discrimination Against Indigenous Populations*—is widely accepted *as a working definition*. This study describes "indigenous communities, peoples, and nations" as, *those that have a historical continuity with 'pre-invasion' and pre-colonial societies that developed on their territories*; further, this study describes "indigenous communities and peoples" *as those that consider themselves as distinct from the other sectors of the societies now prevailing in those countries, or parts of them.*

In the present state of affairs, the "indigenous communities and peoples" form *non-dominant sectors of society,* determined to preserve, develop, and transmit to future generations, *ancestral territories, and their ethnic identities, as the basis of their continued existence as peoples, in accordance with their own cultural pattern, social institutions, and legal systems.* According to this perspective, IK would be the *traditional knowledge of indigenous peoples.* IK is therefore a subset of "traditional knowledge," but not all "traditional knowledge" is indigenous. In other words, *indigenous knowledge is traditional knowledge, but not all traditional knowledge is indigenous.*

Heritage (H)

According to the WIPO, the term, "heritage"[31] is described as follows:

> The *Heritage* of indigenous peoples *includes all moveable cultural property* as defined by the relevant conventions of The United Nations Educational, Scientific and Cultural Organization (UNESCO); *all kinds of* literary and artistic works such as music, dance, song, ceremonies, symbols and designs, narratives, and poetry; *all kinds of* scientific, agricultural, technical, and ecological knowledge, including cultigens, medicines, and the rational use of flora and fauna; human remains; immoveable cultural property such as sacred sites, sites of historical significance, and burials; and documentation of indigenous peoples' *heritage* on film, photographs, videotape, or audiotape.

Thus, H, literally encompasses almost everything.

Interrelationship of H, TK, IK, TCE, and GR

It is important to know that TK has been the focus of the WIPO for a number of years. As per the WIPO's research findings, "traditional knowledge holders," are the ones who create, develop, and practice TK. TK *itself is a subset of* the H. According to the WIPO: "traditional cultural expressions" (TCE) *are a subset of* "traditional knowledge" (TK). Further, "indigenous knowledge" (IK), which is the *"traditional knowledge of indigenous peoples, is also a subset of* 'traditional knowledge' (TK)." As some "traditional cultural expressions" (TCE) are created by indigenous people, *there is an overlap between* TCE and IK, as well. In addition, *some* TK *is closely associated with* GR.[32]

Examples for TK and TCE

The word "traditional" implies "knowledge or an expression" *which has a traditional link with a community*: that it is *developed, sustained, and passed on from one generation to another,* and *carried forward within the community.* In short, what makes that knowledge or expression "traditional" is the relationship with the community. The best way to understand these concepts is to look at some examples.[33a]

TK

Examples of TK include—

- In Thailand, some traditional healers, use the *plao-noi* plant *to treat ulcers;*
- In southern Africa, the San people use—while on hunting expedition—the *hoodia* plant *to put off hunger;*
- In Oman and Yemen, as well as in Iran, *sustainable irrigation is maintained through traditional water systems,* such as the *qanat* the *aflaj,* respectively;
- In the Hudson Bay region, the *Cree Indians* and the *Inuit* maintain unique bodies of knowledge on the *seasonal migration patterns of particular species;* and
- In the western Amazon, indigenous healers use the *Ayahuasca* vine *to prepare various medicines, imbued with sacred properties.*

TCE

Examples of TCE include—

- ***Verbal expressions***: stories, tales, poetry, riddles, signs, elements of languages, such as names, words, symbols, and indications, etc.;
- ***Musical expressions***: songs and instrumental music;
- ***Expressions by actions***: dances, plays, artistic forms of rituals, etc.; whether or not reduced to a material form; and
- ***Tangible expressions***: drawings, paintings, carvings, jewelry, metal ware, textiles, designs, carpets, sculptures, pottery, terracotta, crafts, mosaic, needlework, basket weaving, woodwork, costumes, musical instruments, and architectural forms.

According to the WIPO[33b]:

> For many holders, TK and its form of expression (TCE) are seen *as an inseparable whole.* For example, a traditional tool *may embody* TK *but also may be seen as a* TCE in itself because of its design and ornamentation.

The Intellectual Property System for Protection of TK and TCE

An effective IP system for protecting TK and TCE should consider four dimensions, namely—community, national, regional, and international. Indeed, the overall effectiveness of such a system depends upon the integration and coordination between each level. It is already evident from the above discussion, that *the current conventional system of IP* (which is based on patents, copyrights, trademarks, GIs, etc.) *cannot effectively protect* TK and TCE. Therefore, *alternative approaches*—known as *the positive* and *defensive protections*—may be considered.

Positive Protection[34]

In essence, "positive protection" is the granting of IP rights that: (a) empower communities to promote their TK/TCE; (b) control their uses by third parties; and (c) benefit from their commercial exploitation. This approach enables holders of TK/TCE: (a) to acquire IP rights and assert them, if they needed to; (b) to *prevent* unwanted, unauthorized, or inappropriate uses by third parties (including culturally offensive or demeaning use); and/or (c) *to exploit* TK/TCEs commercially, such as, through the granting of licenses, as a contribution to their economic development.

The options for "positive protection" include:

1. *Existing IP laws* and *legal systems* (including the law of unfair competition);
2. *Extended* or *adapted IP rights* specifically focused on TK (*sui generis* aspects of IP laws);
3. New, stand-alone *sui generis systems* which give rights in TK as such; and
4. Other *non-IP options* can form part of the overall menu, including *trade practices* and *labelling laws*, the *law of civil liability*, the *use of contracts*, *customary* and *indigenous laws* and *protocols*, *regulation of access* to genetic resources and associated TK, and *remedies based on such torts* as *unjust enrichment*, *rights of publicity*, and *blasphemy*.

Defensive Protection[34]

The second approach, "defensive protection," is designed to *prevent the illegitimate acquisition* or *maintaining of* IP *rights* by third parties. Thus, the purpose of the defensive protection is *to stop people outside the community from acquiring* IP *rights over* TK/TCE. For example, India has compiled *a searchable database of traditional medical knowledge* that can be used as *evidence of prior art by patent examiners when assessing patent applications*. Defensive strategies might also be used to protect sacred cultural manifestations, such as *sacred symbols or words, from being registered as trademarks*. Thus, *defensive approach* uses *a range of IP tools* for the protection of TK and TCEs.

Biopiracy, the Misappropriation of TK and Some Remedies[35]

"Biopiracy" is defined *as the process through which the rights of indigenous cultures to genetic resources and knowledge are "erased and replaced for those who have exploited indigenous knowledge and biodiversity.*" In fact, there are many instances of patents obtained on GR and TK, *without the consent of the rightful holders of the resources and knowledge from the developing countries.*

1. The U.S. Patent No. 5,304,718 on *quinoa* (granted to researchers of the Colorado State University), and the U.S. Plant Patent No. 5,751 on *ayahuasca*, a sacred and medicinal plant of the Amazonia, are two excellent examples for this. Similar patenting attempts have occurred on products *based on plant materials and knowledge developed and used by local/indigenous communities*, such as—the *neem tree, kava, barbasco*, and *endod*, among others. Many of these patents have been revoked by the competent national authorities when their validity was challenged in courts;
2. Thus, the *Council of Scientific and Industrial Research* of India, challenged the validity of the U.S. Patent No. 5,401,5041 granted for the *wound healing properties of turmeric*. The USPTO revoked this patent after *ascertaining that there was no novelty, as this has been known in India for centuries (prior art)*;
3. Further, the patent granted to WR Grace Company and the U.S. Department of Agriculture on *neem* (EPO Patent No. 436257) was also revoked by the European Patent Office on the grounds *that its use has been well known in India*;
4. Another patent on *basmati rice* lines and grains (U.S. Patent No. 5,663,484) was also challenged by the *Council of Scientific and Industrial Research*. *Basmati rice* is exclusively grown in the Indian sub-continent, and India is the largest exporter of this product in the world. While the *Council of Scientific and Industrial Research* succeeded in getting these patents revoked, it could not prevent the U.S. company from using the trademark *Texmati* which is deceptively similar to *Basmati*. India learnt a lesson from this experience and the *Agricultural and Processed Food Products Export Development Authority* protected *basmati rice* as a GI for the regions where it is grown in India.
5. Interestingly, while the USPTO *revoked this patent*, it has also made the following observations[36]:

> Informal systems of knowledge often *depend upon face-to-face communication, thereby limiting access to the information to persons in direct contact with one another. The public at large does not benefit from the knowledge nor can the knowledge be built upon. In addition, if information is not written down, that information is completely inaccessible to patent examiners everywhere as prior art when they are examining patent applications.* It is possible, therefore, for a patent to be issued claiming as an invention technology that is known to a particular indigenous community. *The fault lies not with the patent system, however, but with the inaccessibility of the knowledge involved beyond the indigenous community.* The U.S. patent granted for a method

of using turmeric to heal wounds, referred to during India's intervention in June 1999 and again in October 1999, is an example of a patent issued because prior art references were not available to the examiners. In that instance, however, the patent system worked as it should. The patent claim was cancelled based on prior art presents by a party that requested re-examination.

References

1. (a) http://www.wipo.int/edocs/pubdocs/en/intproperty/450/wipo_pub_450.pdf; (b) http://www.ipindia.nic.in/writereaddata/images/pdf/act-of-2000.pdf; (c) https://www.gov.uk/register-a-design; (d) https://www.uspto.gov/patents-getting-started/patent-basics/types-patent-applicationsproceedings; and (e) http://www.china-iprhelpdesk.eu/sites/all/docs/publications/China_IPR_SME_Helpdesk-Design_Patent_Guide.pdf
2. (a) See, Reference 1(a), page 13; (b) https://www.uspto.gov/patents-getting-started/patent-basics/types-patent-applications/design-patent-application-guide; (c) See, Reference 2(b); and (d) See, Reference 2(e), page 1.
3. http://www.wipo.int/designs/en/faq_industrialdesigns.html
4. (a) https://www.uspto.gov/patents-getting-started/patent-basics/types-patent-applications/design-patent-application-guide#def; (b) http://www.china-iprhelpdesk.eu/sites/all/docs/publications/China_IPR_SME_Helpdesk-Design_Patent_Guide.pdf; (c) http://www.ipindia.nic.in/designs.htm; (d) http://www.china-iprhelpdesk.eu/sites/all/docs/publications/China_IPR_SME_Helpdesk-Design_Patent_Guide.pdf; and (e) See, 35 U.S.C. § 173. https://www.uspto.gov/patents-getting-started/patent-basics/types-patent-applications/design-patent-application-guide
5. https://www.statista.com/statistics/276306/global-apple-iphone-sales-since-fiscal-year-2007/
6. In 2011, Procter and Gamble (P&G), the world's largest consumer products company, made 167 international design applications using WIPO's *Hague System for the International Registration of Industrial Designs* and topped the list of applicants, three years in a row. To see an interview/discussion by WIPO magazine of two P&G senior business executives, see: http://www.wipo.int/wipo_magazine/en/2012/06/article_0004.html
7. (a) http://www.wipo.int/treaties/en/registration/hague/summary_hague.html; (b) http://www.wipo.int/treaties/en/classification/locarno/summary_locarno.html; (c) http://www.wipo.int/classifications/locarno/en/; (d) See, *Locarno Agreement*, Article 2(1). http://www.wipo.int/wipolex/en/treaties/text.jsp?file_id=285822; and (e) http://www.wipo.int/treaties/en/convention/summary_wipo_convention.html
8. (a) http://www.wipo.int/sme/en/documents/upov_plant_variety_fulltext.html; (b) http://www.china-iprhelpdesk.eu/sites/all/docs/publications/EN_Plant_Verieties.pdf; (c) http://lawmin.nic.in/ld/P-ACT/2001/The%20Protection%20of%

20Plant%20Varieties%20and%20Farmers%E2%80%99%20Rights%20Act,%202001.pdf; and (d) https://www.uspto.gov/patents-getting-started/patent-basics/types-patent-applications/general-information-about-35-usc-161

9. (a) http://www.upov.int/edocs/pubdocs/en/upov_pub_437.pdf; (b) http://www.wipo.int/sme/en/documents/upov_plant_variety_fulltext.html; (c) See, Reference 8(b); and (d) http://plantauthority.gov.in/about-authority.htm
10. https://www.uspto.gov/patents-getting-started/patent-basics/types-patent-applications/general-information-about-35-usc-161#heading-1
11. (a) http://www.upov.int/export/sites/upov/about/en/pdf/353_upov_report.pdf; (b) See, paragraph 5, http://www.upov.int/edocs/pubdocs/en/upov_pub_437.pdf
12. See, page 2, http://www.upov.int/edocs/pubdocs/en/upov_pub_437.pdf
13. See, Provisions and Limitations: https://www.uspto.gov/patents-getting-started/patent-basics/types-patent-applications/general-information-about-35-usc-161#heading-1
14. (a) http://www.wipo.int/geo_indications/en/faq_geographicalindications.html; (b) http://www.ipindia.nic.in/about-us-gi.htm; and (c) See, page 328, https://www.wto.org/english/docs_e/legal_e/27-trips.pdf
15. http://www.wipo.int/geo_indications/en/faq_geographicalindications.html
16. (a) http://www.teaboard.gov.in/TEABOARDCSM/NQ==; (b) See, page 9, http://www.wipo.int/edocs/pubdocs/en/geographical/952/wipo_pub_952.pdf; (c) https://en.wikipedia.org/wiki/Champagne; (d) https://en.wikipedia.org/wiki/Tequila; (e) https://en.wikipedia.org/wiki/Kondapalli_Toys; and (f) See, page 11, http://www.wipo.int/edocs/pubdocs/en/geographical/952/wipo_pub_952.pdf
17. See, page 13, http://www.wipo.int/edocs/pubdocs/en/geographical/952/wipo_pub_952.pdf
18. https://en.wikipedia.org/wiki/Kondapalli_Toys
19. (a) See, pages 28–32, http://www.wipo.int/geo_indications/en/faq_geographicalindications.html; (b) http://www.wipo.int/edocs/pubdocs/en/geographical/952/wipo_pub_952.pdf; and (c) See, pages 13–14, to understand the terminology, http://www.wipo.int/edocs/pubdocs/en/geographical/952/wipo_pub_952.pdf
20. See, http://www.wipo.int/geo_indications/en/faq_geographicalindications.html
21. http://www.wipo.int/lisbon/en/
22. See, Chapter 5, Section 5.6(2).
23. (a) See, 17 U.S.C. §§ 901–914; (b) https://www.uspto.gov/web/offices/pac/mpep/consolidated_rules.pdf; (c) http://www.wipo.int/wipolex/en/treaties/text.jsp?file_id=294976; (d) https://www.wto.org/english/docs_e/legal_e/27-trips.pdf; See, Section 6, Articles 35–38; and (e) The *Washington Treaty*, which provides protection for the ICLDs, has not yet entered into force, but has been ratified/acceded to by the following States: *Bosnia and Herzegovina, Egypt and Saint Lucia.*
24. (a) See, *Manual of Patent Examining Procedure* (MPEP), § 150.1: https://www.uspto.gov/web/offices/pac/mpep/consolidated_rules.pdf; (b) http://sicldr.gov.in/Resources/SICLD-Act-English.pdf; and (c) http://english.sipo.gov.cn/laws/lawsregulations/200804/t20080416_380325.html
25. http://sicldr.gov.in/faq

26. (a) https://www.copyright.gov/title17/92chap9.html; (b) https://www.uspto.gov/web/offices/pac/mpep/consolidated_rules.pdf; and (c) http://www.wipo.int/wipolex/en/details.jsp?id=12739
27. http://www.wipo.int/edocs/mdocs/tk/en/wipo_grtkf_ic_22/wipo_grtkf_ic_22_inf_8.pdf
28. (a) See, page 25 (Terminology), http://www.wipo.int/edocs/pubdocs/en/tk/768/wipo_pub_768.pdf; (b) See, pages 42–43, http://www.wipo.int/edocs/mdocs/tk/en/wipo_grtkf_ic_22/wipo_grtkf_ic_22_inf_8.pdf; (c) See, WIPO/GRTKF/IC/13/7 (2008): Inter-Governmental Committee (IGC) on Intellectual Property and Genetic Resources, Traditional Knowledge and Folklore, Annex, page 5 (2008).
29. (a) See, pages 17–18, http://www.wipo.int/edocs/mdocs/tk/en/wipo_grtkf_ic_22/wipo_grtkf_ic_22_inf_8.pdf; (b) https://www.cbd.int/doc/legal/cbd-en.pdf; (c) http://www.wipo.int/wipolex/en/details.jsp?id=9446; and (d) http://www.fao.org/plant-treaty/overview/en/
30. (a) See, page 40, http://www.wipo.int/edocs/mdocs/tk/en/wipo_grtkf_ic_22/wipo_grtkf_ic_22_inf_8.pdf; and (b) Article 1, "Draft Articles on Traditional Cultural Expressions as Prepared at IGC 19 (July18–22, 2011)," as incorporated in document "Matters Concerning the Intergovernmental Committee on Intellectual Property and Genetic Resources, Traditional Knowledge and Folklore (IGC)" (WO/GA/40/7). Article 2 "Beneficiaries" of the "Draft Articles on Traditional Cultural Expressions as Prepared at IGC 19 (July18–22, 2011)," as incorporated in document "Matters Concerning the Intergovernmental Committee on Intellectual Property and Genetic Resources, Traditional Knowledge and Folklore (IGC)" (WO/GA/40/7).
31. See, page 23, http://www.wipo.int/edocs/pubdocs/en/tk/768/wipo_pub_768.pdf
32. See, page 26, http://www.wipo.int/edocs/pubdocs/en/tk/768/wipo_pub_768.pdf
33. (a) See, pages 14–16, http://www.wipo.int/edocs/pubdocs/en/tk/933/wipo_pub_933.pdf; and (b) See, page 13, http://www.wipo.int/edocs/pubdocs/en/tk/933/wipo_pub_933.pdf
34. See, page 22, http://www.wipo.int/edocs/pubdocs/en/tk/933/wipo_pub_933.pdf
35. Correa, C. M. (2001). *Issues and Options Surrounding the Protection of Traditional Knowledge, A Discussion Paper.* Geneva, Switzerland: Quaker United Nations Office.
36. See, page 7, http://www.tansey.org.uk/docs/tk-colourfinal.pdf

7

Trade Secrets

If you want to *keep a secret*, you must also *hide it from yourself.*

George Orwell

"Trade secret" (TS) is a form of intellectual property (IP) of a company—encompassing manufacturing or industrial secrets (such as *formulae, patterns, compilations, programs, devices, manufacturing methods, techniques, or processes*)[1a] and commercial secrets (such as *strategic cost structure, knowledge of consumer profiles, customer buying preferences and requirements, pricing strategies, advertising, sales and distribution strategies*, etc.)[1b]—*that provide competitive advantage to it in the market place.*[1c,d] Unlike other forms of intellectual property, trade secrets are *a maintain-it-yourself form of protection*. A TS is not registered with the government and so a company must do everything it can to keep it confidential. Thus, TS protection lasts only as long as the secret can be kept confidential.

The Definition

According to the *Uniform Trade Secrets Act* (UTSA)[2a]:

> Trade secret, means information—*including a formula, pattern, compilation, program, device, method, technique, or process* that:
>
> a. *Derives independent economic value, actual or potential, from not being generally known to, and not being readily ascertainable by proper means* by, other persons who can obtain economic value from its *disclosure* or use; and
> b. *Is the subject of efforts that are reasonable under the circumstances to maintain its secrecy.*

The Evolution of Trade Secret Law

It is important to note that there is *no statutory definition* for a TS in the United States. Thus, for a long time, TS protection was handled *exclusively under the jurisdiction of state law*. In 1985, the *National Conference of*

Commissioners approved and recommended for enactment in all the states, the UTSA *with* 1985 *Amendments*,[2a] in Minneapolis, Minnesota.

As shown above, the UTSA defined TS and codified the common laws relating to TS. Another significant development is the *Economic Espionage Act* (EEA) *of 1996* (18 U.S.C. §§ 1831–1839),[2b] which made the *theft or misappropriation of a trade secret a federal crime.* This law has two provisions: the first, 18 U.S.C. § 1831(a), criminalizes *the theft of trade secrets to benefit foreign powers.* The second, 18 U.S.C. § 1832, criminalizes *this theft for commercial or economic purposes* (different statutory penalties apply to the two crimes). However, in spite of this federal law, and similar criminal statutes legislated in various states, a large number of TS misappropriation cases were still settled in civil courts under state law. Subsequently, 48 states in the United States of America (USA) modeled their statutes after the UTSA and made an effort to harmonize them. On the other hand, the states that did not adopt the UTSA (example, New York and Massachusetts), use their own state statutes and/or continue to apply common law.[3]

Consequently, as *there are still real differences* in how states define TS and set the scope of TS protection, one may find that courts in different states reach different or even contrary conclusions regarding the protection of a TS. Nevertheless, the United States (U.S.) courts *have generally found a variety of technical and non-technical information to be* TS, such as formulae, recipes, customer lists, business know-how, internal machines/methods of production, and blueprints. Meanwhile, what is also clear, is that the U.S. Courts *do not afford* TS *protection to common information* or *procedures*, such as—the *lemonade recipes, cooking procedures for barbeque chicken, and customer lists posted on a company website.*

Even if the information encompasses "failed attempts," it may still receive protection as a TS. For instance, *information on failed attempts and painful losses*—such as the details of Research & Development (R&D) efforts that are unsuccessful in circumventing the scale-up issues in the manufacture of products, R&D dead-ends, abandoned innovations due to prohibitively high costs, or the unsuccessful attempts to increase sales—*is eligible to receive protection as a* TS. That is because even such "negative" information could be highly valuable, as it can teach a competitor what not to do and prevent unproductive investments.

Commenting about the TSs, World Intellectual Property Organization states[4]:

> While these conditions vary from country to country, *some general standards exist* which are referred to in Article 39 of the *Agreement on Trade-Related Aspects of Intellectual Property Rights*:
>
> a. The *information must be secret* (i.e., it is *not generally known among, or readily accessible to,* circles that normally deal with the kind of information in question);

b. It *must have commercial value* because it is a secret; and
c. It *must have been subject to reasonable steps by the rightful holder* of the information *to keep it secret* (e.g., through confidentiality agreements).

Differences between a Patent and Trade Secret

There are major differences between a patent and TS. Patent application procedures are complex and involve statutory guidelines and government approval. On the other hand, trade secrets are protected without registration, and without even procedural formalities, such as the disclosure of information to a government authority. Further, while the subject matter of patents *must be novel*, a trade secret *need not be novel*. In fact, a trade secret that is *not easily decipherable* to the competitors may still contain pieces of information that are available from public sources. Patent protection (the grant of an exclusive right) exists *for a limited period of time* (in exchange for public disclosure of the invention), after which the *information is available for use in the public domain*. On the contrary, TS *protection exists indefinitely*, as long as the TS holder can successfully maintain its secrecy. Patents grant the inventor *exclusive exclusionary rights*, while TS laws *do not grant the holder any exclusive rights*. In fact, competitors who can *independently and lawfully decipher* a TS through reverse engineering are allowed to commercially exploit it. If commercial exploitation of a TS is not possible without revealing the critical information, then patenting and/or copyright protection is advisable. On the contrary, if a company can effectively safeguard the critical information, and it is not easily decipherable by others, then, TS protection is advisable. It is *more difficult to enforce* a TS than a patent. In fact, *the protection given to a* TS *varies significantly from one country to another*, and *is considered weak*, compared to *patent protection, which is uniformly strong all over the world*.

Misappropriation of TS

In a competitive business world, it is not uncommon for companies to spy on each other *using lawful means* to *learn about each other's trade secrets and/or neutralize the competitive advantage* (through reverse engineering or employee poaching). Unfortunately, in competitive fields, companies may sometimes resort to *unlawful methods* such as *industrial espionage*. Thus, if a TS is *acquired by industrial espionage, it is deemed to be misappropriated*. Fortunately, *acts of industrial espionage are deemed illegal under* the EEA of 1996.[2b] Consequently, the acquirer of TS will be subject to *legal liability for having acquired it improperly*. On the other hand, the TS holder must prove that: (a) the information has *independent commercial value (actual/potential)* and (b) reasonable steps were taken *to safeguard information against espionage.*

TS Protection by Courts—Enforcement

In general, the U.S. courts are willing and committed to enforce the TS protection. In fact, courts take TS *misappropriation claims* seriously and investigate them diligently, because they wish to: (a) protect the rights of a company to safeguard its critical information; (b) not handicap an individual's ability to work in a chosen industry; and (c) not restrict the use of information to a particular industry. Therefore, the courts try to balance the *right of a company to protect its critical information* against *the rights of* others *to use readily available data* or *expertise* gained through long employment in a specific industry.

Normally, companies try to protect their *confidential proprietary information* in two ways: through *non-compete agreements* (NCAs)[1] and *non-disclosure agreements* (NDAs).[2] Interestingly, however, courts will not hesitate to invalidate any *overreaching non-compete agreements.* Grounded in realism, courts will nullify NCAs *if the terms are unreasonable in scope, duration, and territory.* For example, an NCA that takes away the ability of a former employee to work for a competitor for 10 years, will be almost certainly invalidated. On the other hand, an NCA that prohibits a former employee (who had direct access to a TS) from working for a competitor for 1 year may be potentially enforceable. It is also noteworthy that the enforceability of TS agreements will depend on the jurisdiction. For example, NCAs can be either severely restricted or even barred in California, under certain circumstances. This is because California courts require proof that an NCA is absolutely vital to protect a TS.

Proving a TS Violation in Court—Procedure

With regards to proving a violation of TS, courts do not insist on a magic bullet formula, because there is none. Instead, they broadly examine the circumstances surrounding the specific case and determine whether the TS owner

1 A non-compete agreement restricts the ability of a former employee or partner to work for a direct competitor or otherwise compete with the company, for a specified period of time. The rationale for this type of agreement is that the information will retain its value for a limited period of time.

2 A non-disclosure agreement is the second type of agreement that prevents a former employee or partner to disclose a company's sensitive, commercially valuable information.

employed reasonable measures to safeguard the TS and whether the violation took place in spite of them. Therefore, to prove a TS violation in court, the TS owner must produce the following evidence[5]:

1. That the TS has *independent economic value* (actual/potential);
2. That *reasonable measures* have been taken *to protect the secrecy of* TS;
3. That the TS is *not easily decipherable;*
4. That the accused *has access* to the TS; and
5. That the accused *has been notified (namely is aware) of the confidentiality* of TS.

To start with, the economic value of a TS can be established in terms of *increased revenues, profits, or even market share.* In addition, the economic value of a TS can be shown in terms of the *enhanced competitive positioning made possible because of the* TS, which effectively diminishes the ability of a competitor to compete against the TS owner. Secondly, the TS owner need not show that drastic measures were taken to prevent the access to and dissemination of the TS; instead, the TS owner has to only demonstrate that *reasonable measures were taken* to protect the TS. In this regard, the courts look for any evidence, such as—the presence or lack of fences, general access restrictions, locked doors, warning sign postings, and visitor access restrictions. Thirdly, the TS owner has to show that: (a) the accused was a former employee; (b) the accused had access to the TS that is neither publicly known nor readily accessible; and (c) the TS was clearly marked, "confidential/proprietary."

Proven TS Violations—Remedies

The TS owners have multiple remedies available once the TS violation has been proven in court, including—*injunctions, lost profits* (calculating these requires a historical baseline), *unjust enrichment, reasonable royalties and punitive damages, plus attorneys' fees* (willful and malicious acts). In addition, *criminal remedies* are also available for *industrial espionage.* The penalties for such federal crimes include fines of up to U.S. $10,000,000 for companies and a fine of U.S. $500,000 coupled with a jail sentence of up to 15 years for individuals. Indeed, there have been multiple convictions under the *EEA 1996.*

It is important to know that in cases of TS violation, several types of injunctions are available to the TS holder.[6] First, immediately upon commencement of the lawsuit, the TS holder may seek *temporary injunctive relief* (or *temporary restraining order*) due to the imminent risk of loss of the TS. A *temporary injunctive relief* is limited in scope, generally in effect until a hearing can be held on a request for a *preliminary injunction.* The *preliminary injunction's* goal is to prevent further use/disclosure of the TS while the case is pending. To obtain a *preliminary injunction,* the presentation of evidence and witness testimony is required, much like a trial. Following the trial, a *final injunction* barring the use/disclosure of the TS may be issued. The defendant, however, remains able to seek termination of the *final injunction* when the TS becomes generally known through legitimate means. Perhaps it is strategically in the best interest of the TS owner to consolidate the *preliminary injunction hearing* with *the trial* on the liability case, having the issues of damages and punitive remedies postponed to a separate trial. Thus, when the court considers *injunctive relief as inappropriate,* it may direct the misappropriator to pay the TS holder an ongoing royalty.

The TS owner faces a dilemma in this regard: to prove a TS violation, the TS owner must first disclose the TS information during the litigation. Fortunately, in a TS case, a *protective order* can be issued early by the court which: (a) *limits* the disclosure of the TS information *only to a small group of people* and (b) *restricts* the use of the TS information *only for the purpose of the lawsuit.* Further, at the end of the law suit, *the protective order* requires the *return/destruction of the TS information.*

Strategic Decision by Companies—Patenting versus TS Protection

While *protecting critical information that gives a company competitive advantage*—such as formulae, patterns, compilations, programs, machines, devices, manufacturing methods, techniques, or processes—*it must make a strategic decision*: "whether to seek patent protection" or "maintain the information as a trade secret." Indeed, this decision involves a consideration of multiple issues such as[3a]:

1. What is the company's *present competitive ranking* (in *revenues, profits,* and *market share*);
2. What is the *level of competition* in the market place;
3. What is the *nature of the subject matter* to be protected;
4. Is the information *not readily available* and *not easily accessible;*
5. Does the information have *real/potential economic value;*
6. Does the information *give the company sustainable competitive advantage;*
7. Is it *easily decipherable* (reverse engineered);
8. Can the information be *independently developed* (time, effort, funding, facilities);
9. Is the subject matter *patentable;* and
10. What does it take to *maintain the information as a* TS?

Based on the answers to the above questions (and perhaps others!), a company will have to make the *strategic decision* between *patenting* and *protecting the information as a* TS. In this regard, several points are worthy of mention. First of all, it may not be possible to protect the information as a patent, because it may not meet the conditions of either *novelty* or *non-obviousness*. On the other hand, one must surely consider *patent protection* when the information to be protected involves—*novel technical innovations or improvements to inventions*. Secondly, one must also consider the *differences in the scope of protection* between the two forms of IP. A TS *does not prevent competitors* from reverse engineering or *independently developing the same invention* or product technology (thus, a TS affords *weak protection*). On the other hand, a *patent right excludes others* from making, using, or selling the invention within the United States, as well as *prevents others* from importing the invention into the United States. Consequently, it would be considered as infringement when someone who independently develops or reverse engineers a patented invention (thus, a patent affords *strong protection*). Thirdly, because the patent information is readily available, competitors can guess what a company's strategic goals are and *may attempt to circumvent the patented technology*. A TS *keeps the competition guessing*. Fourthly, *the duration of protection* plays a decisive role. A TS may be able to *protect the information indefinitely*, as long as its secrecy is maintained, while patent *protection only lasts for a limited period of time*. Finally, one must also *consider the costs involved* in patenting versus TS protection, before choosing *one type of protection over the other*. Clearly, the bottom line here is that there are no simple pathways that everyone can follow. These decisions depend upon *the company, customers, competition, and the market*, to say the least. Therefore, the right approach is to consider all aspects and choose the form of protection that best suits the information to be protected and the strategic goals of the business.

Development of a Strong TS Policy—Consistent Measures

To develop a strong TS policy, one must do a thorough analysis of various issues involved. There are many instances, in which a company loses its TS because it does not have a strategic TS policy in place. The most important aspect of protecting a TS is *knowing how to protect a* TS *through reasonable and consistent measures*. One may think that this shouldn't be too difficult to do, but *flawless and consistent protection of information as a secret, while still making use of the information* (TS) *in the business, is indeed quite challenging*. The following is *a short list of consistent measures* that may be considered by *companies seeking effective and sustainable protection of* TSs[7]:

1. ***Employee policies and practices from entry to exit***. These include protocols for—new employee interviews, new employee orientation programs, employee joining agreements, employee exit interviews,

and employee exit agreements. This involves many important steps taken from the beginning to the end of an employee's stay with the company, such as—(a) *interviews* to get to know the employee better, and inform him/her about the TS policy; (b) *employee orientation programs* that inform and educate employees regarding what can and cannot be discussed or shared outside the company; (c) *strict access policies and documentation methods* to guard the confidential proprietary information as secret (example, NDAs); (d) *employee-exit interviews and reminders* regarding TSs; and (e) *employee exit agreements* (example, NCAs);

2. ***Screening procedures***. Strict procedures must be put in place *to screen all the materials that the company wishes to disclose to the public or third parties* (investors, shareholders, vendors, suppliers, buyers, contractors, etc.)—including, sales or advertising materials, product launch details, publications, or speeches—to make sure that confidential proprietary information is not accidentally leaked out;
3. ***Security practices***. These include: (a) *creating secure access and control* for TS information (such as by requiring employees to use multi-level password protection, enforcing secure data storage, transmission and disposal mechanisms, and using restricted access to limit the physical access to a TS); (b) *ensuring notice of confidential proprietary information* (such as by labeling/marking all materials, products, and documents as CONFIDENTIAL); (c) *placing security signs and reminders*; and (d) *periodically reviewing the security policies and procedures* to ensure compliance throughout the company;
4. ***Licensing methods***. This includes thorough periodic appraisals of all the confidential user licenses (in and out), intranet user screen notifications, and disclaimers by the company. In addition, the company should be able to confirm that all its obligations to maintain third-party licensed TSs are fully met and that licensees are fulfilling their obligations to maintain the TS; and
5. ***External surveillance***. Companies should have policies for external surveillance and conduct systematic and periodic external monitoring of the marketplace to ensure that there is no unauthorized use or misappropriation of company's TSs.

Representative Examples for TS

TSs include *two types of confidential proprietary information of companies*, namely—manufacturing or industrial secrets (such as formulae, recipes, patterns, compilations, algorithms, programs, devices, manufacturing methods, techniques, or processes) and commercial secrets (such as strategic cost structure, knowledge of consumer profiles, customer buying preferences and requirements, supplier contracts, pricing strategies, advertising, sales, and distribution strategies, etc.)[1b]—*that offer competitive advantage to the companies in the marketplace.* Thus, *trade secrets encompass many categories of subject matter that are industry-specific and company-specific.*

As *trade secrets* are indeed "actual secrets" of companies, and companies go to great lengths to safeguard those secrets, it is almost impossible to know "real world TS examples" in every single category of subject matter listed above. Consequently, we outline a few well-known examples for TSs[8a,b]:

1. Coca-Cola formula
2. Kentucky Fried Chicken formula
3. Lena Blackburne's Baseball Rubbing Mud
4. NY Times Best Seller list (algorithm)
5. Google Internet engine's search algorithm
6. WD-40 formula
7. Twinkies formula
8. Big Mac sauce recipe
9. Krispy Kreme donuts recipe

Coca-Cola. The *Coca-Cola* story is instructive (source: Tracy Wilson, "How *Coca-Cola* Works?")[9a] to understand that *trade secrets are not born, but built through a lot of hard work*. Thus, the TS story of *Coca-Cola* comprises of—unique product, smart and courageous decision to protect the information as a *trade secret* as opposed to protecting it as a *patent*, extraordinary, yet sustainable measures to protect the *trade secret*, understanding of the evolving consumer perceptions and preferences, and developing business strategies that culminate in competitive advantages. The *Coca-Cola* story started in 1886, when Dr. John Stith Pemberton of Jacob's Pharmacy in Atlanta, GA, invented a "magical" syrup and mixed it with carbonated water to create the world's most-famous drink, known today as *Coca-Cola*. Dr. Pemberton marketed his cola innovation first as a "health tonic," and sold the drink for 5 cents/glass. Next, Dr. Pemberton worked with Frank Robinson, who named the drink, and suggested it in cursive handwriting, as *Coca-Cola*.[9b] This trademark created the *brand identity* for the cola, which became increasingly resilient over 130 years. Today, *Coca-Cola* is acknowledged by everyone as one of the most recognizable brands in the entire world. It is safe to say that no one has ever been able to reproduce the exact *Coca-Cola* formula since 1886.

According to the *Cable News Network*[9c]:

> The *original recipe was only written down in* 1919, more than half a century after a reported morphine addict and pharmacist John Pemberton invented the drink in 1886. Until then, *it was passed down by word of mouth.*
>
> The formula was finally committed to paper when a group of investors led by Ernest Woodruff took out a loan to purchase the company in 1919. "As collateral, he *provided a written record of the Coca Cola secret formula,*" Coke said in a statement on its site.
>
> Since the 1920s, *the document sat locked in a bank in Atlanta,* until Coca-Cola decided to emphasize the secret in its marketing strategy. 86 years later, Coca-Cola *moved the recipe into a purpose-built vault* within the World of Coca-Cola, the company's museum in Atlanta. The *ambiance is made complete by red lighting and fake smoke.*
>
> Coca-Cola has always claimed *only two senior executives know the formula at any given time,* although they have never revealed names or positions. But according to an advertising campaign based around the recipe, *they can't travel on the same plane.* The *vault,* like one straight from a film, *has a palm scanner, a numerical code pad, and massive steel door.*

KFC. Next, the story of *Kentucky Fried Chicken* (KFC) is also very interesting, as it teaches several important aspects of *trade secrets* (source: McManus, M. R. "KFC Recipe").[10a] Thus, in 1930, Colonel Sanders invented a tasty recipe for the coating of fried chicken made from 11 herbs and spices and started to sell it in his roadside restaurant in North Corbin, Kentucky. Colonel Sanders successfully franchised his recipe to his friend Harman, and thus the first KFC Franchise was born in Utah in 1952. KFC popularized its brand of chicken in the fast food industry by challenging the dominance of the hamburger. The way KFC protected its TS is again instructive. Initially, Sanders *withheld the secret recipe to himself in his head,* though *he eventually wrote it all down.* Thus, his *original, handwritten copy of the KFC recipe is hidden in a safe* in Kentucky, and *only a few select employees, bound by a confidentiality contract, know the exact recipe.* For further protection, *two separate companies each blend a portion of the mixture,* which is then *run through a computer processing system to standardize its blending* (source: KFC.com).[10b] That same recipe is still used today in all the KFC restaurants. According to rumors, *the employees who know the recipe can't ever travel together by plane or auto*; to safeguard the secret further, KFC *modernized its security systems,* the *recipe was temporarily moved to another secret, secure location* via an armored car, which was further *guarded by a high-security motorcade.* Today, the fact that there are over 20,500 KFC outlets in more than 125 countries and territories around the world attests to KFC's success as a global brand.

Lena Blackburne's Baseball Rubbing Mud. In 1938, Lena Blackburne, an athletics coach, *invented a secret recipe* (the "baseball rubbing mud") from special mud and water *that can dull the surface of glossy new baseballs, and create an easier grip, without wrecking the balls*[11a] (source: Bintliff, "Joy in Mudville").[11b] Blackburne also ensured that no baseball rules were broken. Prior to this, some baseball umpires also attempted to use shoe polish, tobacco juice, and the dirt beneath their feet to similarly fix the balls. Unfortunately, none of these attempts succeeded. This is so because, while the substances dulled the ball surfaces, they also damaged the baseballs in the process (source: Bintliff).[11b]

The specialty of Blackburne's recipe—crafted from rich mud found in southern New Jersey near the Delaware River—is that *it didn't wreck the balls.* As soon as the Athletics' chief umpire gave a green signal, the *American League* teams began clamoring for Lena Blackburne's Baseball Rubbing Mud. Soon, the *National League* teams followed suit, and this made Lena Blackburne's Baseball Rubbing Mud very famous (source: Bintliff).[11b]

According to Bintliff[11b]:

> The mud is on public land, but *we've always kept the location a secret* to keep people from trampling it. I make about five or six trips to the mud hole a year. When I get the mud to my house, I rinse it with tap water and filter out debris. I'd like to expand the business. Recently I sold a few buckets to a few NFL teams, *which have found that the mud makes footballs easier to grip.* We're hoping to hear from more of them.

NY Times Best Seller List. In October 1931, the NY Times unpretentiously published, for the first time, the *Best Seller List* was for the books sold just in New York City, comprising of five fiction titles and four non-fiction titles (source: Miller, L. "The Best Seller List as Marketing Tool and Historical Fiction").[12a,b] This *Best Seller List* was then expanded to eight cities in the following month, with a separate list published for each city. The list grew gradually to include 14 major cities in the United States of America, and by April 1942, the NY Times added the *Sunday Book Review section* and included, for the first time, The *National Best Seller List.* This list included 17 fiction titles and 16 non-fiction titles, based on the number of cities that ranked them as their best sellers. Subsequently, the NY Times eliminated the *city lists* and focused only on the *National Best Seller List,* which was based on reports from leading booksellers in 22 major cities. By 1950s, many professionals began to use the *National Best Seller List* of the NY Times as the list to monitor the best-selling books, in conjunction with the *Publishers Weekly.* In 1960s and 1970s, several retail bookstores (such as B. Dalton, Crown Books, and Waldenbooks) introduced a new a business model of selling newly published books *that made the NY Times Best Sellers List.*[12b] By 2004, the *NY Times Best Seller List* included data collected from 4,000 bookstores, as well as an unspecified number of wholesalers. By 2011, NY Times published two new e-book lists, the first to track the sales of combined print and e-books, the second to track only the e-book sales (each of these lists are further sub-divided into fiction and non-fiction). In December 2012, the NY Times divided the children's books list into two new lists: *middle-grade* (ages 8–12) and *young adult* (ages 12–18), both of which include *book sales across different platforms* (hard, paper, and e-book).[12b]

It must be noted that the exact methodology by which the NY Times compiles the *National Best Seller List* is a TS, and it is an *evolution-in-progress.*

Thus, every single example of the TSs listed above share three aspects in common. They are:

1. Independent economic value;
2. Reasonable measures in place to protect the secrecy of TS; and
3. Not easily decipherable.

As a result, the TS holder enjoys *competitive advantages* for an *indefinite period of time*—often, for periods much longer than patent protection (>20 years)—at least, until the *trade secret* is deciphered or "somehow" leaks into the public domain.

References

1. (a) https://www.uspto.gov/patents-getting-started/international-protection/trade-secret-policy; (b) http://www.wipo.int/sme/en/ip_business/trade_secrets/trade_secrets.htm; (c) https://nelligan.ca/article/trade-secrets-what-are-they-and-how-can-you-protect-them-from-unauthorized-disclosure-and-use/; and (d) https://definitions.uslegal.com/p/proprietary-information/
2. (a) http://www.uniformlaws.org/shared/docs/trade%20secrets/utsa_final_85.pdf; and (b) https://www.gpo.gov/fdsys/pkg/PLAW-104publ294/pdf/PLAW-104publ294.pdf
3. (a) See, the article by Dawn Rudenko Albert of Dickstein Shapiro LLP, New York. http://www.iam-media.com/Magazine/Issue/42/Management-report/Trade-secrets-in-the-United-States; and (b) http://www.beckreedriden.com/trade-secrets-laws-and-the-utsa-a-50-state-and-federal-law-survey-chart/
4. http://www.wipo.int/sme/en/ip_business/trade_secrets/protection.htm

5. See, the article by Dawn Rudenko Albert of Dickstein Shapiro LLP, New York. http://www.iam-media.com/Magazine/Issue/42/Management-report/Trade-secrets-in-the-United-States
6. http://www.ipo.org/wp-content/uploads/2013/04/IP_Protection_for_Trade_Secrets_and_Know-how1076598753.pdf
7. See, page 96, http://www.iam-media.com/Magazine/Issue/42/Management-report/Trade-secrets-in-the-United-States
8. (a) https://money.howstuffworks.com/10-trade-secrets10.htm; and (b) https://info.vethanlaw.com/blog/trade-secrets-10-of-the-most-famous-examples
9. (a) https://money.howstuffworks.com/coca-cola.htm; (b) http://www.coca-colacompany.com/stories/coke-lore-trademark-chronology; and (c) See, Ivana Kottasova's article: http://edition.cnn.com/2014/02/18/business/coca-cola-secret-formula/index.html
10. (a) See, Melanie McManus's article: https://money.howstuffworks.com/10-trade-secrets9.htm; and (b) https://www.kfc.com/menu/promotions/10-chicken-share
11. (a) https://money.howstuffworks.com/10-trade-secrets1.htm; and (b) See, Jim Bintliff's article: http://money.cnn.com/2008/08/19/smallbusiness/lena_blackburne_mud.fsb/index.htm
12. (a) Miller, L. (2000). The best seller list as marketing tool and historical fiction, page 286, in *Book History*, Ed. Greenspan, E. and Rose, J., Vol. 3; and (b) https://en.wikipedia.org/wiki/The_New_York_Times_Best_Seller_list

Section III

Strategic Intellectual Property Management

8

Intellectual Property (IP) Management Strategies and Tactics

Strategy is a *pattern* in a *stream of decisions*.

Henry Mintzberg

According to the World Intellectual Property Organization (WIPO),[1]

> "*Intellectual Property* (IP) refers to creations of the mind, such as inventions, literary and artistic works, designs and symbols, names, and images used in commerce." "IP is needed for many reasons: first, *the progress and well-being of humanity rest* on its capacity to create and invent new works in the areas of technology and culture. Second, the legal protection of new creations *encourages the commitment of additional resources for further innovation*. Third, the promotion and protection of intellectual property *spurs economic growth, creates new jobs and industries*, and *enhances the quality and enjoyment of life*."

Thus, the progress and well-being of societies and nations depends upon *how well their needs, wants, challenges, and opportunities are addressed by innovations and how effectively the innovations are protected by IPRs*. According to this author: "innovations address the societal needs, wants, challenges, and opportunities—by questioning status quo and creatively leveraging ideas and resources—to create and capture value." Consequently, IP *management strategies and tactics* are important for *value creation, value capture, and promotion of innovations*.

In addition, in today's world, the *market valuation of companies is exceptionally dependent upon the intellectual capital (IC) of the firms*. Not surprisingly, therefore, organizations are forced to reckon with the twin needs of "being continually innovative" and "having effective IP management strategies and tactics in place"—*to build strong competitive advantage, achieve sustainable economic growth, and enhance the value of the company*.

Sources and Types of Innovation

There are two *sources of innovation*: *technology* or *business model* (See Figure 8.1). Figure 8.1 shows how *technology* and *business model* determine the type of innovation.

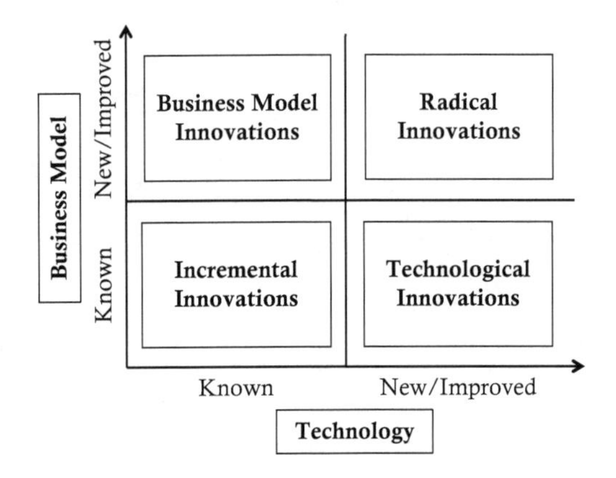

FIGURE 8.1
The sources and types of innovations.

Accordingly, there are four *types of innovations* possible[1]:

1. *Incremental innovations*: these innovations use *known* technologies and *known* business models;
2. *Technological innovations*: these innovations use *new/improved* technologies, but *known* business models;
3. *Business model innovations*: these innovations use *new/improved* business models, but *known* technologies; and
4. *Radical innovations*: these innovations involve both *new/improved* technologies and *new/improved* business models.

Typically, resilient firms target a strategic combination of all four types of innovations in their innovation portfolios. In addition, they employ "IP management strategies and tactics" *to build competitive advantage to achieve sustainable economic growth.*

[1] In addition, we find *Breakthrough* and *Disruptive Innovations* in the management literature. They both apply to *Technological* and *Business Model* Innovations. *Breakthrough Innovations* improve an existing market, while *Disruptive Innovations* disrupt the existing markets and create new markets. See, Christensen, C. M., *The Innovator's Dilemma*, Harper Business (2011).

Forces of Innovation

The *forces of innovation* are the very *reasons* why companies innovate. They are *five-fold* (See Figure 8.2):

1. **Demands for growth and expansion**. Companies depend on shareholders and investors for raising capital to achieve growth and expansion. In turn, investors and shareholders expect to see continually growing returns on their investments. Thus, *demands for sustainable growth and expansion* by investors and shareholders force companies *to be continually innovative;*
2. **Evolving customer requirements**. Companies understand that the needs/wants/challenges of customers continually evolve, and that their long-term market success depends upon their *ability to meet them faster and better than competition*. Thus, *evolving customer requirements* force companies *to be continually innovative;*
3. **Market changes**. In the commercial world, there is only one thing that a company can be sure of, market changes. These can take any form, such as—a new market entrant introducing a competitive product, an existing rival launching a superior product, and a supplier relationship or a buyer relationship that is increasingly less profitable, and the like. Consequently, companies must not only possess the ability to anticipate, but also the ability to effectively respond to *market changes*. Thus, *market changes* force companies *to be continually innovative.*

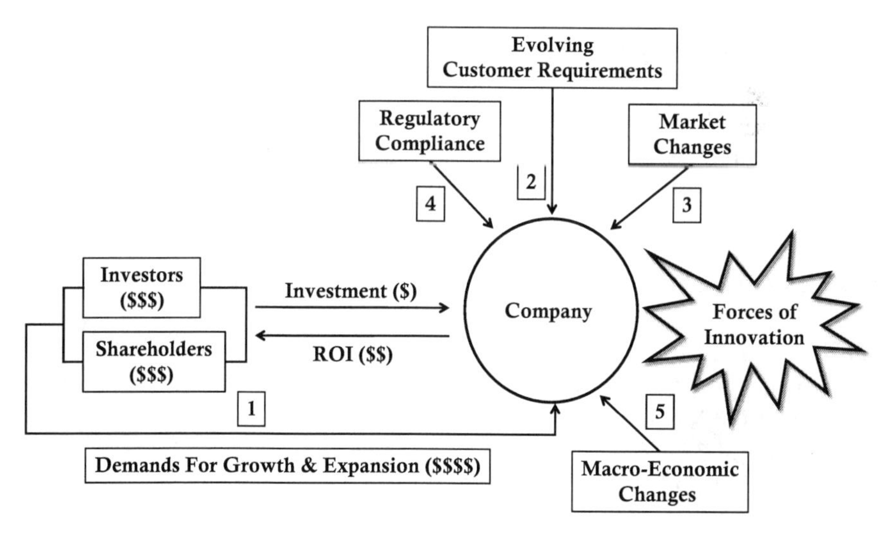

FIGURE 8.2
The forces of innovation.

4. **Growing regulatory compliance standards**. Today, in many instances, consumer protection groups and environmental watchdog organizations are successful in raising customer awareness of the products and business practices of companies and informing them about any safety and environmental concerns. As a result, customers are becoming sophisticated and demand solutions that not only address their needs, wants, or challenges, but also meet the social and environmental standards. As a result, policy makers—at the county, state, and federal levels—are continually bringing stronger regulations that demand higher social/environmental standards from the companies. Thus, growing *regulatory compliance standards* force the companies *to be continually innovative.*
5. **Macro-economic changes**. Even when all is well with the market performance of a company, unexpected national/international events, such as—major natural calamities, economic crises resulting from business corruption, terrorist attacks, political instability, and wars—could significantly deteriorate the macro-economic situation and dramatically impact the profit/loss picture of companies. Thus, unexpected *macro-economic changes* force companies to *be continually innovative.*

As elaborated above, companies must be "innovative" in order to: (a) capitalize on the *sources of innovation*; (b) take advantage of the *forces of innovation*; and (c) achieve *competitive advantage and sustainable economic growth.* In order to do so, companies must understand the relationship between *strategy, innovation,* and IPRs—all designed for *competitive advantage.*

Linkage between Strategy, Innovation, and IPRs

There is a general myth that *all patents promote innovation.* This is certainly not true. In an earlier publication, the "IsIPR Model," has shown that[2]:

- All IPRs *may not promote innovation*; and
- Only IPRs *protecting sustainable innovations promote innovation.*

Here, we extend this line of thinking further. Thus, companies wishing to develop *effective strategies for* "IP protection" and "IP management" must understand the linkage between *innovation, IPRs,* and *strategy for competitive advantage.* Figure 8.3 shows that *commercially successful companies,* in general, possess three things in common: (a) they each have (some) *strategy for competitive advantage* regardless of how effective it might be; (b) they each (may) possess at least some *IPRs*; and (c) they all are *innovative,* though the extent may differ.

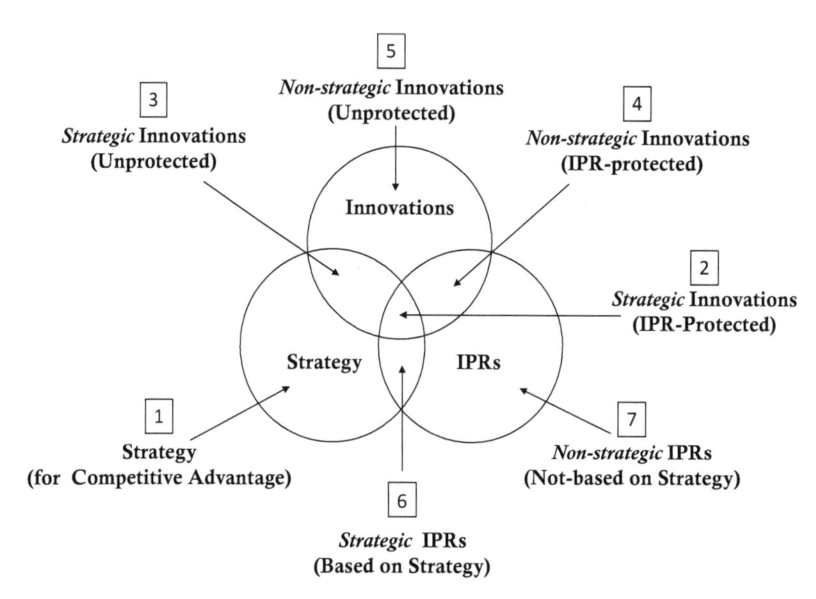

FIGURE 8.3
The linkage between innovation, strategy, and IPRs.

Thus, companies wishing to develop *effective strategies for* "IP protection" and "IP management" must leverage the following *determinants of competitive advantage*:

1. Based on understanding of the market place[3] (customer, competition, suppliers, buyers, industry structure, etc.) and evaluation of one's own core competencies,[4] every company must develop its own *strategy for competitive advantage* (see 1 in Figure 8.3). Only those companies that have developed a clear *vision* and *mission* can also have an effective *strategy for competitive advantage*;
2. *Strategic* innovations can be of two kinds: "IPR-protected" *strategic* innovations (2 in Figure 8.3) and "unprotected" *strategic* innovations (3 in Figure 8.3). Among the two, the "IPR-protected" *strategic* innovations may help the company *achieve long-term competitive advantage*. Normally, "unprotected" *strategic* innovations provide "only temporary" *competitive advantage* and are likely to *lose to imitations, potentially creating big losses for the company*. However, "unprotected" *strategic* innovations (3 in Figure 8.3) that are *not easily decipherable* and safeguarded as *trade secrets*—may provide *long-term competitive advantage*;
3. Companies engaged in *strategic* innovations (see 2 and 3 in Figure 8.3) will have greater *competitive advantage* than those engaged in *non-strategic* innovations (see 4 and 5 in Figure 8.3) which are of two kinds: "IPR-protected" *non-strategic* innovations (4 in Figure 8.3) and

"unprotected" *non-strategic* innovations (5 in Figure 8.3). If not properly managed,[1] "IPR-protected" *non-strategic* innovations may cost the company unproductive IP maintenance fees, while "unprotected" *non-strategic* innovations pose *infringement risks* and *decisional predicaments* for *new product development;*

4. Companies may possess two kinds of IPRs: *strategic* IPRs (based on strategy) (see 6 in Figure 8.3) and *non-strategic* IPRs (not-based on strategy) (see 7 in Figure 8.3). Companies view *strategic* IPRs (6 in Figure 8.3) as high priorities for innovations that can afford *competitive advantage,* while *non-strategic* IPRs (7 in Figure 8.3) often will not make this cut. Consequently, *non-strategic* IPRs may end up costing the company *if they are not properly managed as part of company's comprehensive IC management*[2]; and
5. "IPR-protected" *strategic* innovations (see 2 in Figure 8.3) *do promote innovations,* while "IPR-protected" *non-strategic* innovations (see 4 in Figure 8.3) *do not promote innovation,* as long as they not a part of company's business strategy. However, only those companies that are adept in *comprehensive IC management* know how to convert these IP assets into profitable assets and create a revenue stream through *sale, license, or joint venture creation.*

Companies Desire for Competitive Advantage

Companies strive for *competitive advantage* for multiple reasons:

- To grow the *revenues, profits, and market share of the company;*
- To provide *value to customers* (vendors, suppliers, buyers, etc.);
- To provide *value to shareholders and investors;* and
- To achieve *sustainable growth and expansion.*

Thus, companies work very hard to gain *competitive advantage* to create a *win* for company, *win* for customers, and *win* for shareholders and investors. This depends upon how well the company manages its *IC* and *intellectual assets.* Figure 8.4 clarifies the importance of *strategic IP management.*

Thus, *strategic IP management* is critical for many reasons: (1) it identifies and protects all *strategic intellectual property* (otherwise, companies may fail to adequately protect them); (2) it helps to create "IP-based" products and services that are *differentiated;* (3) it creates "IP-protected" *strategic* innovations that offer *competitive advantage;* (4) it allows companies to market

[1] Comprehensive *intellectual capital management* will solve this problem.
[2] Will be elaborated in the next chapter.

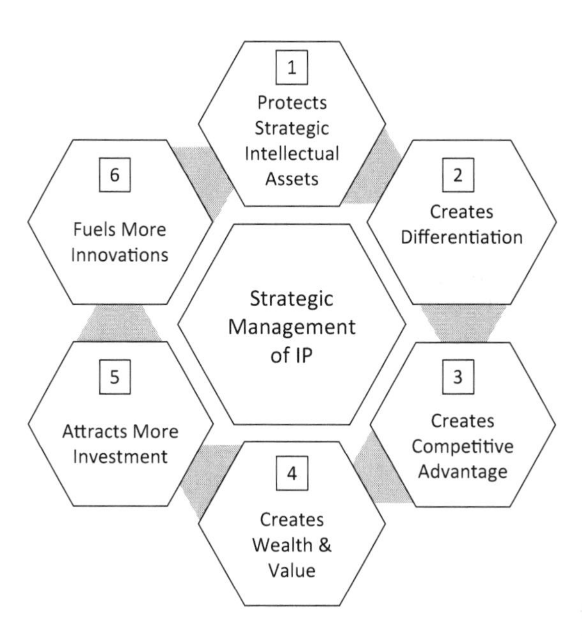

FIGURE 8.4
The value of strategic management of IP.

differentiated products and services and *make stronger claims*, while preventing the competition from easily copying them—that contributes to *wealth and value creation*; (5) companies having expertise in strategic management of intellectual property attract *more investments for growth and expansion*; and (6) more investments *fuel more innovations*, and the innovation cycle continues. *Strategic IP management* requires *dynamic planning, implementation, evaluation, and adjustment of IP strategy*—based on clear understanding of the customer, competition, market, investment climate, and innovation opportunities by the firms.

Strategic IP Management, Competitive Advantage, and Sustainable Growth

In this regard, it is instructive to note what Bill Gates of Microsoft Corporation says about *Strategic IP Management*[5a]:

> "Over the last 10 years, it has become imperative for CEOs to have not just a general understanding of the intellectual property issues facing their business and their industry, but to have quite a refined expertise relating to those issues... It is no longer simply the legal department's problem.

> CEOs must now be able to formulate strategies that capitalize on and maximize the value of their company's intellectual property assets to drive growth, innovation, and cooperative relationships with other companies."
>
> **Microsoft Chairman, Bill Gates 2004**

For a firm to effectively manage its IP assets, it must know how *strategic IP management* is planned and executed. Two decades ago, it was not uncommon to find firms that treated IP protection of inventions and innovations as "costs" on their balance sheets, not anymore. As Bill Gates rightly pointed out, today, only companies that know how to strategically protect, manage, and leverage the IP asset will be successful *in achieving competitive advantage and sustainable economic growth in the industry.*

Thus, *strategic IP management,* requires a company to be adept at the following (Figure 8.3):

1. Design *innovation strategy as part of business strategy* to build *strategic* IP assets;
2. Execute *innovation strategy* to integrate discovery, invention, and innovation;
3. Create "wealth" (revenues, profits, and market share) from "IPR-protected" *strategic* innovations and *strategic* IPRs (see Figure 8.3) *through commercialization;*
4. Leverage "IPR-protected" *non-strategic innovations* and *non-strategic IPRs* into revenues *through licensing, sale, or joint venture creation;*
5. Safeguard "unprotected" *strategic* innovations using *innovative defensive strategies* (to be elaborated shortly);
6. Divest or donate "unprotected" *non-strategic* innovations (see Figure 8.3);
7. While steps 1–6 are important to a company in the *strategic management* of "internal" IP assets, a company may sometimes require IP from *outside the company.* For this reason, companies must also be good at *identifying* "external" IP assets *that fit the business strategy* and *acquiring them through in-licensing, purchase, or strategic alliance;*
8. Create "value" by *managing the IP assets as part of the IC of the firm*[3] and *ensuring that stakeholders are informed, as the valuation of the firm depends on it;*

---

[3] Thomas Stewart defined IC as, "the intellectual material—knowledge, information, intellectual property, experience—that can be put to use to create wealth; the intellectual material that has been formalized, captured, and leveraged to create wealth by producing a higher-valued asset; it is packaged useful knowledge." See Chapter 1.

9. Conduct a *periodic audit of the IP assets of the company to determine what needs to be done* in terms of steps 3–8 discussed above[5b]; and
10. Monitor *the market place and the IP landscape of the company* periodically and systematically *to identify infringements* and be prepared to *enforce IP rights when necessary.*

Indeed, *strategic IP management* is complex and requires knowledge and expertise in many aspects of the company, such as—company's vision and strategic priorities, technology and innovation, market place (customer, competition, vendors, and suppliers, etc.), and legal opportunities.[6] Consequently, *strategic IP management* in companies is done by a core team of senior business experts from research & development (R&D), marketing, and legal functions.[4] Therefore, *strategic IP management* offers *competitive advantages* to a firm among its rivals in the industry, and this ensures its *sustainable growth.*

Approaches for Strategic IP Management

According to the WIPO,[7] companies may employ a combination of *strategic approaches for IP management.* Figure 8.5 summarizes these approaches:

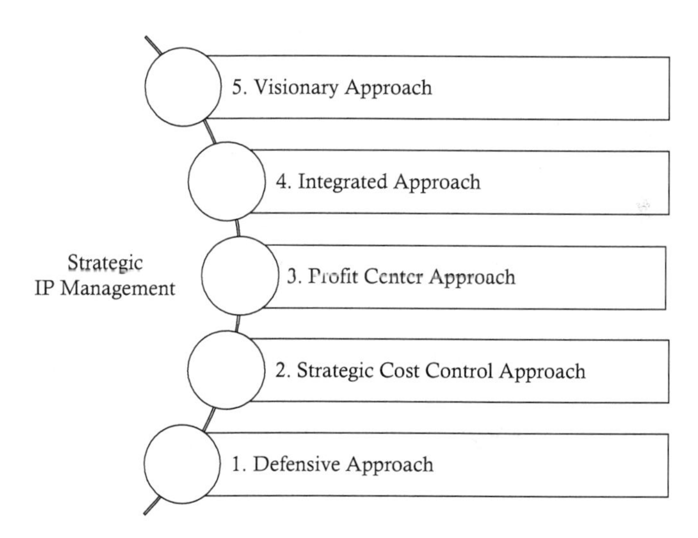

FIGURE 8.5
Strategic IP management approaches adopted by companies.

[4] In many organizations, the R&D, strategy, and legal functions are poorly integrated. As a consequence, firms miss opportunities to create and exploit the value of intellectual property. See reference 6.

The details are as follows[7]:

1. ***Defensive approach.*** The purpose of this approach is *to fortify the company's business through a defensive* IP *shield by safeguarding it from external* IP *threats and "*IP *infringement risks."* Thus, a company engaging in *strategic IP management* must occasionally do IP landscaping and *ensure that its* IP *has not been infringed by anyone.* Next, a company *must aggressively deal with "*IP *infringement risks" by enforcing its* IP *rights,* through litigation and/or out-of-court settlements, *as necessary.* In addition, IP landscaping enables a company *to identify "external* IP *threats."* Indeed, these can be converted into opportunities for expansion and growth by the company, through—*in-licensing, purchase of* IP, *cross-licensing,* or *strategic alliances;*
2. ***Strategic cost control approach.*** The purpose of this approach is *to strategically control the costs of filing and maintenance of* IP, as opposed to blindly filing and maintaining IP, without paying any attention to its fit with company's strategic vision and mission. This requires a company to fully align its *IP strategy* with its own *vision, mission,* and *business strategy.* When a company follows the *strategic cost control approach,* it will have an opportunity *to invest in and retain all "strategic* IPRs", *while minimizing the costs of "non-strategic* IPRs" (see Figure 8.3);
3. ***Profit center approach.*** The purpose of this approach is *to find strategic opportunities to convert* "non-strategic IPRs" *into revenue stream* (see section above, step 4). In this context, the opportunities for the company include—*out-licensing, sale, cross-licensing,* or *strategic alliances.* Interestingly, *this may provide a platform* for companies to *discover adjacent/new business opportunities for profitable expansion and growth;*
4. ***Integrated approach.*** The purpose of this approach is *to strategically expand the value of IPRs to the entire company—i.e., all the functional units of the company*—by aligning the *IP strategy* with company's *business strategy.* This can happen when the company ensures that its *strategic innovations* and *strategic* IPRs (2 and 6, see Figure 8.3) are specifically designed to take into consideration the "larger interests" of multiple functional units. What this means is that the *competitive rivals to the company cannot easily copy* the IPRs, as they are *complex* and *integrated.* Thus, the company holding *strategic innovations* and *strategic* IPRs *gains competitive advantage over its rivals;* and
5. ***Visionary approach.*** The purpose of this approach is to broaden the vision of companies for sustainable development. Accordingly, companies that *invent technologies and business models to deliver sustainable economic, social, and environmental value* are most likely to gain *competitive advantage and succeed* in the future. This gives such companies opportunities *to create IP that is state-of-the-art and superior and achieve sustainable growth.*

Thus, it is clear that companies looking to gain *competitive advantage* and *sustainable growth* should be committed to *strategic IP management* and utilize a *combination of the IP approaches* that fit their vision, mission, and business strategy.

Defensive and Offensive Patent Strategies

In the context of *strategic IP management*, it is important to point out the strategies that *IP-savvy companies* use for protecting arguably the most important form of IP, namely—*patents*.

1. ***Defensive patent strategies***. These approaches for *patent protection*, include—*in-licensing, purchase, cross-license, or strategic alliance*. In addition, companies use a particular approach known as *building a patent wall* to protect *a specific patent of business interest* (Figure 8.6);

   This can be achieved by companies in two ways:

   a. Use own R&D to *develop a critical set of invention patents to surround the specific patent of commercial interest*, rather than leave the patent exposed to *external IP threats*; and

   b. Alternatively, *protection of the specific patent of commercial interest* can also be done by *acquiring the necessary patents if available*.

   Indeed, Gillette Corporation presents the best example for the *defensive strategies approach*. Commenting on the role played by defensive strategies in *strategic IP management*, Racherla reported the following[8]:

> Recently a paper was presented at the DRUID in Copenhagen (Sternitzke 2012) that researched the Gillette Corporation, a company known for

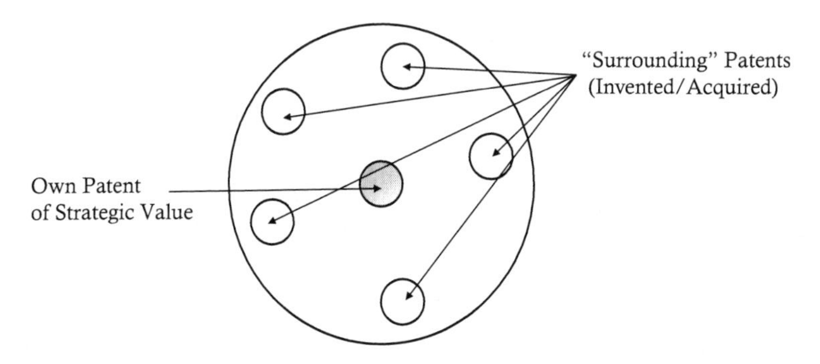

FIGURE 8.6
"Building a patent wall" around an own patent of strategic value.

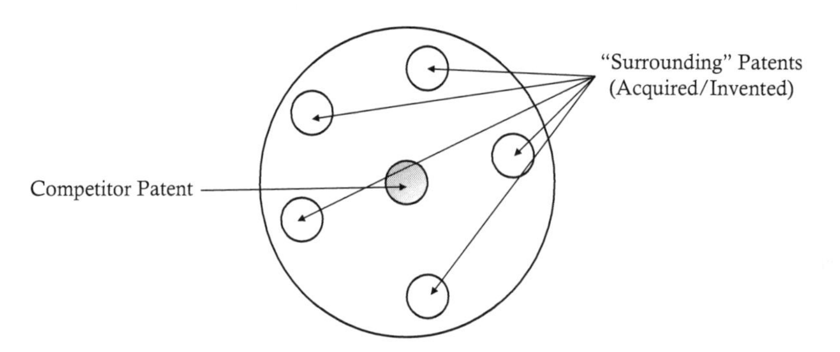

FIGURE 8.7
"Building a picket fence" to diminish/disable a competitor patent.

strategic IP management, and reported how a firm could enhance value appropriation of modular and integral products by creating interlocking IP. Thus, the study examined patent filings on the Gillette Fusion razor and concluded: "Several levels of interlocking activities were identified, including patent claims following a hierarchical structure and filings of overlapping content in multiple patent families. In general, the degree of coupling of technological components as well as business model achieves the creation of interlocking patents which, for integral systems, further enhances complexity." It is precisely this complexity resulting from interlocking patents that protects innovations of the Gillette Corporation (now P&G), builds sustainable competitive advantage, and promotes future innovations; and

2. ***Offensive patent strategies***. This kind of approach for *patent protection* includes—*detection of infringements, aggressive enforcement of patent rights (litigation), and settlement of patent battles outside of court.* The outcomes may include stopping the infringer from manufacturing the infringed product, preventing the sale of infringed product, getting the infringed product removed from the market place, and collecting compensatory fines and punitive damages. In addition, *companies are legally allowed to build a picket-fence of patents around a competitor's patent to diminish/disable the invention* (Figure 8.7). This is similar to *building patent wall to debilitate the competitive edge of a rival.*

Non-practicing Entities or Patent Trolls

"Non-practicing entities" (NPEs)—otherwise known as, "patent assertion entities" or "patent trolls"—*do not actually produce or sell goods, but work incessantly to amass profits through IP litigation and licensing.*[9] In recent years, they have been notoriously attracting significant scrutiny in the IP system of the United States of America. As President Obama noted[10]:

> "[NPEs] don't actually produce anything themselves. They're just trying to hijack somebody else's idea and see if they can extort some money out of them."

The recognition of patent assertion entities *negative role in the innovation ecosystem* by the United States Congress resulted in many unsuccessful attempts to pass legislation to curb their activities, such as *the Strong Patent Act* and *the Innovation Act*. However, inability to come up with a *universally acceptable definition* for NPEs and their activities, prevented such legislation from passing. Indeed, the following aspects of NPEs/patent assertion entities from the *patent assertion and US Innovation report*[10] by the Obama White House are worth noting:

1. NPEs *do not practice their patents*; further, they *do not conduct* R&D on any *technology/products* related to their patents;
2. NPEs *do not engage in technology transfer* (transforming the patent language into a usable product or process);
3. NPEs *operate as predators,* they often *wait until someone sinks huge investments into an innovation and then move on them to assert their claims;*
4. NPEs acquire patents *solely for the purpose of extracting payments from alleged infringers;*
5. NPE's litigation strategies *capitalize on their non-practicing status, which insulates them to counter-claims of patent infringement;*
6. NPE's *acquire patents that make vague claims*[5] and *use them to litigate several companies at once for infringement;* further, they *demand moderate license fees from all companies at once,* presenting little/no specific evidence of infringement to back their litigation, *hoping that at least some companies* (who made irrecoverable investments) *will settle the lawsuits instead of risking expensive, long trials with uncertain outcome;*
7. NPEs *conceal their identity by creating numerous shell companies* and require those who settle to sign non-disclosure agreements; this disables defendant's ability to create common defensive strategies (such as sharing of legal fees rather than settling individually). Thus, NPEs *often mask their identities, acquire patents with vague claims, some of even questionable validity,* and *assert them to extract settlement fees.*

[5] In the United States, NPEs purchase and assert patents that were granted by a specific set of examiners, who tend to allow incremental patents with vaguely worded claims. See reference 9(b) for details.

NPEs could be of four different types[9a]:

a. Shell companies *which purchase controversial patents from others only for asserting them;*
b. Companies *that originally sold products,* but have either completely or partially closed their operations;
c. Intermediary agents *that assert patents* on behalf of patent owners; and
d. Law firms *that help clients to exploit their IP,* for a contingency fee.

Litigation by NPEs and Impact on Innovation and R&D

To understand how NPEs actually operate and benefit from their litigations, the example of Ray Niro is illustrative. Thus, Ray Niro, who is dubbed as *the first patent troll* by his own admission, sued 235 companies and made $315,000,000 in 18 years of patent litigation.[11]

According to some observers[9a]:

> The *strategy* taken by patent trolls *slows the progress of science* in several ways. First, patent trolls *increase the transactional costs* associated with developing technology and with claiming IP rights. In a legal system where patent trolls thrive, the development of a new technology must be accompanied by extensive searches for related patents. Not only must well known patents of competitors be taken into account, but obscure patents with only vague connections to the technology being developed must be identified. *These searches can add considerable cost and reduce the likelihood that new technologies will be developed.* The legal costs associated with litigation initiated by patent trolls also *reduce the funding available for innovation,* and these legal costs can rise to quite significant amounts. Personnel who would otherwise be engaged in promoting innovation throughout the organization *will have their attention diverted elsewhere by the litigation, which will consume the human resources of the technology developers in addition to the financial resources.*

The Harvard-Stanford study[9b] found that the *issuance of ill-defined patents* by the patent examiners and the *inconsistencies in the clarification of claims across patent examiners* aids the rent-seeking behavior of NPEs and increases the likelihood of litigations among practicing entities. In the European Union, NPEs face different challenges: the litigation fees are lower (*incentive*), while the losing party is required to bear the costs of litigation for both parties (*disincentive*).

Given the circumstances, *what can companies proactively do to prevent attack by NPEs and minimize the litigation costs due to NPEs?* Though there is no foolproof way to avoid NPEs, *strategic IP management* which takes into consideration the following, may be helpful:

1. Periodic IP landscaping to make sure that company's *strategic* innovations and *strategic* IPRs do not infringe on other's IPRs;
2. Buy, license, or cross-license IP to "build a patent wall" around company's *strategic* innovations and *strategic* IPRs; and
3. Invent *inter-locking patents* (company's own R&D may have to do this) to "build a patent wall" around company's *strategic* innovations and *strategic* IPRs.

References

1. http://www.wipo.int/edocs/pubdocs/en/intproperty/450/wipo_pub_450.pdf
2. Racherla, U. S. (2016). Do IPRs promote innovation? in *Innovation and IPRs in China and India, Myths, Realities and Opportunities*, Eds. Liu, K.-C. and Racherla, U. S., Springer, Singapore.
3. Porter, M. (1979). How competitive forces shape strategy, *Harvard Business Review*, pp. 137–145.
4. Prahalad, C. K., Hamel, G. (1990). Core competence of the corporation, *Harvard Business Review*, pp. 79–90.
5. (a) http://www.hgf.com/services/strategy/; (b) http://www.goodwinprocter.com/~/media/Files/Publications/Attorney%20Articles/2003/Key_Strategies_for_Successful_Intellectual_Property_Asset_Management_Part_1_of_3.pdf
6. http://www.hbs.edu/faculty/Publication%20Files/CMR5504_10_Fisher_III_7bbf941f-fe1b-4069-a609-9c6cd9a8783b.pdf
7. http://www.wipo.int/edocs/mdocs/innovation/en/wipo_inv_mty_02/wipo_inv_mty_02_5.pdf
8. Racherla, U. S. (2016). Do IPRs promote innovation? in *Innovation and IPRs in India and China: Myths, Realities and Opportunities*, Eds. Liu, K.-C. and Racherla, U. S., China-EU Law Series, Springer.
9. (a) https://www.nalsar.ac.in/IJIPL/Files/Archives/Volume%201/5.pdf; (b) https://scholar.harvard.edu/files/xavier/files/trolls.pdf
10. https://obamawhitehouse.archives.gov/sites/default/files/docs/patent_report.pdf
11. Xenia, P. Extreme makeover: From patent troll to the Belle of the Ball. Retrieved on March 3, 2007 from: http://www.law.com/jsp/article.jsp?id=1171360978084

9

Intellectual Property Valuation

Managers and investors alike must understand that *accounting numbers* are the *beginning, not the end,* of *business valuation.*

Warren Buffet

Elsten and Hill conducted a study on the "intangible asset market value" by examining specifically the role intangible assets in *corporate market capitalization across a range of indices* around the world and showed that companies that invest in *intangible assets* (ITAs) *continually outperform* the S&P 500. Thus, this study showed that *the share of ITAs in the S&P 500's market value* has steadily grown from 17% (in 1975) to 85% (in 2015) (Table 9.1)[1]:

Thus, the last decade witnessed a remarkable growth in the number of companies which attained market leadership through *strategic intellectual property (IP) management*, specifically engaging in the *creation, capture,* and *leverage of* IP. Nevertheless, the value of IP to business is still poorly understood in many companies. In general, small and medium enterprises—the building blocks of rapid economic growth in emerging economies—are particularly slow to realize the potential of *strategic IP management.*

Goals for Intellectual Property Valuation

IP valuation is sought by companies for many different reasons[2]:

1. **To bolster business strategy (buying/selling, creation of joint ventures, mergers, and acquisitions, initial public offering, and filing bankruptcy)**: IP is a key component of the *intellectual capital* of

TABLE 9.1

S&P Market Capitalization of Firms

	S&P 500 Market Capitalization	
Year	**TAs (%)**	**ITAs (%)**
1975	83	17
2015	15	85

the firm and a major determinant of a company's value. An accurate IP valuation is a must for buying or selling a company, creation of a joint venture, mergers and acquisitions, initial public offerings, and filing bankruptcy; in every such transaction, multiple stakeholders base their decisions on the IP valuation. In short, an accurate valuation of IP *fortifies the business strategy of the company;*

2. **To sell, buy, in-license, and out-license IP**: When a company decides to sell or buy IP, it must know what its IP is worth. Similarly, when a company decides to in-license other's IP/out-license its own IP, both parties must be clear about the value of IP that is involved, in order to negotiate the transaction;
3. **To raise capital (debt/equity)**: Companies need to raise capital in order to implement their business strategies and meet their growth plans and expansion targets. Consequently, many IP-intensive companies *use their own IP* (their strength) *as collateral to raise capital*—debt capital (bank loans, bonds, etc.), or equity capital (selling company stock to shareholders/investors). However, it is important to note that banks are somewhat reluctant to accept the risk associated with such transactions using IP assets, as there are no universally accepted and precise IP valuation methods. However, everyone agrees that in the near future, the use of IP as collateral for raising capital will become commonplace, as the standards for IP valuation become more universal. In some cases, *IP-backed securitization is used to raise capital;*[3]
4. **For taxation purpose**: Companies wish to know the value of their IP as precisely as possible, for the purpose of *tax compliance* and *tax deductions;*
5. **To inform external stakeholders of the financial health of the company**: *Financial accounting* and *management accounting* are both important to a business, though each has a different objective. *Management accounting* is useful to business executives *to make decisions concerning the day-to-day operations of a business.* On the other hand, *financial accounting* is useful *to convey the financial health of a company to its external stakeholders.* The audience for *financial accounting* reports are—the board of directors, financial institutions, shareholders, and investors. Consequently, companies must know the *value of their IP* to convey it to all key stakeholders. Thus, publicly traded companies make *financial accounting reports* available on a quarterly basis;
6. **For litigation or dispute resolution**: Knowing the value of IP (as precisely as possible) is critical to litigate IP infringers, or deal with breach of a contract; and
7. **To make effective day-to-day operational decisions**: Managers of the firm who make *internal* operational business decisions, do

so based on *management accounting*. It is important to know that *management accounting* is *helpful to exploit current and future market opportunities*. Indeed, managers often *have to make day-to-day operational decisions in a fluctuating business environment under severe time constraints and based on insufficient data*. Consequently, *management accounting* relies on current and future market IP opportunities. Thus, decisions involving Research & Development (R&D) and commercialization require understanding of the entire patent landscape and implementing company's IP strategy (buy, sell, license/cross-license) as part of its business strategy. Thus, knowing the value of IP is critical to ensure company's financial well-being.

Strategies for Successful Intellectual Asset Management

A firm's *intellectual assets* (IAs)[1] are a sum of the company's intellectual property rights (IPRs), *trade secrets* (TSs), and *Codified Proprietary Knowledge* (CPK).[2] Therefore, a firm's *intellectual asset management* (IAM)[3,4] requires management of three forms of IP, namely—IPRs, TSs, and CPK. IAM is a systematic approach for *leveraging* IAs *and extracting value out of them*. Generally speaking, successful companies follow the steps noted below for IAM:

1. Identify the IAs to *extract value later*;
2. Protect the IAs to *retain proprietary rights to these assets*;
3. Leverage the IAs, especially *trade secrets* and CPK to generate *wealth* and *value* (revenues, profits, market share, and *competitive intangibles, CIs*);[4] and
4. Create new IAs from existing IAs.

[1] See Figure 1.2, Chapter 1.

[2] A firm's CPK includes *all confidential information that is valuable and exclusive property of the company*, such as—standard operating procedure manuals, internal reports that capture their specialist know-how, in-licensing/out-licensing agreements, franchisee contracts, strategic alliances, research notebooks in which secret formulae and unique compositions are noted, and other difficult-to-imitate business/technology processes, among others (see Chapter 1).

[3] *Intellectual capital management* is purposely not taken up in this discussion, as that is meant for senior level business executives and not critical for practicing scientists and engineers (see Chapter 1).

[4] A firm's CIs include *valuable, difficult-to-quantify ITAs, such as firm's culture, innovative abilities, core competencies, unique buyer/supplier relationships, brand equity, reputation, and goodwill*, to mention a few (see Chapter 1).

The details of these steps are as follows:

1. **Identify the IAs**. Audit the *IAs* of the company to differentiate the *non-obvious* IAs from the *obvious* IAs. Generally speaking, when a company does its IA audit, it may *discover unexpected* IAs, in addition to IPRs and *protectable innovations*. For instance, a company may find that its most valuable IAs are in the form of CPK—such as *internal reports that capture their unique specialist know-how, in-licensing/out-licensing agreements, franchisee contracts, strategic alliances, research notebooks in which secret formulae and unique compositions are noted*, and other *difficult-to-imitate business/technology processes*. Sometimes, IAs may be more identifiable by talking to external contractors and consultants, as they look at company's assets in an unbiased way. Furthermore, in the process of identifying IAs and extracting value out of them, a company often finds opportunities to create new IAs;
2. **Protect the IAs**. To reap the benefits of IAs, ensuring protection of IAs is absolutely critical. Thus, companies must find the best possible ways to guard IPRs, TSs, and CPK. Companies need to protect IPRs by using *strategic IP management* methods (see Chapter 8). Similarly, TSs, which are valuable IAs to any company, can yield benefits only if appropriate and reasonable measures are taken to protect such information as secret (see Chapter 7). Finally, CPK must also be protected by labelling it CONFIDENTIAL, giving restricted access only to those who use it, taking precautionary measures (such as Non-Disclosure Agreements [NDAs]). This will then allow companies to *transform valuable information into* "wealth" and "value;"
3. **Leverage the IAs**. IPRs, combined with TSs and CPK, constitute the IAs. A company must *strategically extract their full* "value;" and

   This needs the company to execute *strategic IP management*:

   a. **Defensive approach**. Fortify the company business with a defensive IP shield by safeguarding it from "external IP threats" and "IP infringement risks." Also, deal with "IP infringement risks" by enforcing IP rights, through litigation and/or out-of-court settlements, as necessary. In simple terms, a company must *safeguard* the "differentiation" of its products/services;

   b. **Strategic cost control approach**. Fully align *IP strategy* with own *vision, mission*, and *business strategy*. When a company follows the *strategic cost control approach*, it will have an opportunity to invest in and retain all "strategic IPRs," while minimizing the costs of "non-strategic IPRs" (see Chapter 8, Figure 8.3);

   c. **Profit center approach**. Strategically find ways to convert "non-strategic IPRs" into revenue (see Chapter 8). In this context, the opportunities for the company include—out-licensing, sale, cross-licensing, or strategic alliances;

d. **Integrated approach**. Strategically *expand the value of IPRs to the entire company—i.e., all the functional units of the company—* by aligning the IP strategy with company's business strategy. This can happen when the company ensures that its strategic innovations and strategic IPRs (see Chapter 8, Figure 8.3) are specifically designed to take into consideration the "larger interests" of multiple functional units; and

e. **Visionary approach**. Broaden the vision of company for sustainable development. Accordingly, companies that invent technologies and business models to deliver sustainable economic, social, and environmental value are more likely to achieve competitive advantage and growth.

The decision makers of a company *must be fully integrated into the dynamic context, so they can strategically lead and manage IP—to meet the company's vision, mission, and objectives.*

Some of the questions that they have to grapple with in IAM are:

i. What is the *best course for protecting* a particular IA—*patenting* or *protecting it as a trade secret;*

ii. Should a particular invention be *patented immediately* or should the R&D wait and invent more so that the company can *patent an area* (see *Building a Patent Wall,* Chapter 8) to achieve better defenses;

iii. What is the best way to protect the CPK? Which parts of CPK must be *carefully safeguarded as* TSs;

iv. What is the best way to leverage IPRs for *competitive advantage? New product development? Out-licensing? Sale? Cross-licensing? Joint Venture (JV);*

v. What is the best way to leverage TSs for *competitive advantage;*

vi. What is the best way to leverage CPK for *competitive advantage;* and

vii. How can IAs be strategically leveraged to achieve *high company valuation?*

4. **Create new IAs**. A company *that knows how to execute* steps 1–3 *well* would be in an ideal position for *creating new* IAs.[5] This may seem somewhat counter-intuitive after a discussion of *the identification, protection, and leverage of* IAs. However, IAM is a *strategic management*

[5] See Chapter 8.

approach for: (a) converting IAs *into wealth and value* and (b) transformation of *useful information* into IAs. Steps 1–3 describe how to convert IAs into *wealth and value.* However, they do not teach a company *how to achieve transformation of useful information into* IAs. This requires a company to the following well:

a. Ensure a clear *vision, mission,* and *values;*
b. Set the *business strategy aligned with vision, mission, and values to achieve sustainable competitive advantage*—based on understanding of customer, competition, market opportunities, and own capabilities;
c. Ensure that all employees are *knowledgeable and fully aligned with company's strategic vision, mission, values, and business objectives;* and
d. Create "easy-to-implement" business processes and manage them *to transform all useful information into* IAs, such as—
    - Train employees to create *standard operating procedure manuals* and *internal reports to capture all specialist know-how and difficult-to-imitate business processes;*
    - Standardize *in-licensing/out-licensing agreements* and *franchisee contracts;*
    - Implement *strategic alliances when possible for competitive advantage;* and
    - Develop a *strong documentation archives to store and retrieve research notebooks in which secret formulae and unique compositions are noted.*

The Strategic IA Audit of Firms

Broadly speaking, the purpose of a *strategic IA audit* is to enable a company to take an inventory of its IAs and analyze their fit *vis-à-vis* company's vision, mission, and business objectives—*to identify future opportunities for competitive advantage* and *growth.*

Accordingly, a *strategic IA audit* may be divided into three phases[5]:

1. **The management phase**—wherein an inventory of the IAs is taken *to understand their fit with company's vision, mission, and business objectives;*
2. **The benchmarking phase**—wherein the IA portfolio of the company is *benchmarked against the competitive IA landscape to identify* IA *strengths and weaknesses;*

3. **The opportunity identification phase**—wherein the *future opportunities for IA portfolio are decided in terms of in-licensing, out-licensing, sale, purchase* or *strategic alliance.*

   Each phase of the *strategic IA audit* has different goals and actionable business plans. From a company point of view, the purpose of *strategic IA audit* is to[5]:

   a. Take *stock of the inventory of* IAs—i.e., to understand the portfolio of IAs and IP rights owned by the company;
   b. Figure out the *strengths and weaknesses* of the portfolio of IAs;
   c. Uncover *opportunities and threats* to the company's business;
   d. Objectively assess *the programs, projects, resources, technologies, and business models used for creation, protection, and extraction of* IAs (including those that are owned, used ,and/or acquired);
   e. Enforce IA *ownership rights,* and *deal with infringements, if there are any;*
   f. Identify *under-utilized* and/or *un-utilized* IAs vis-à-vis business strategy;
   g. Explore *revenue generation opportunities using under-utilized* and/ or *un-utilized* IAs (out-licensing, cross-licensing, sale, or strategic alliance); and
   h. Enable *strategic planning* to *create competitive advantage* and *facilitate growth.*

Goals for Strategic IA Audit

A company usually undertakes *strategic* IA audit when it wants to[5]:

1. Know *the valuation of its* IAs (IPRs, TSs, and CPK);
2. Sell/out-license *un-utilized* or *under-utilized* IPRs or form a strategic alliance;
3. Acquire/merge with an IPR-centric company;
4. Improve the *procedures for* IA *protection and management;*
5. Ensure *statutory compliance in a changing legal environment;*
6. Litigate IPR *infringements*/resolve IPR *disputes;*
7. Learn about IP *opportunities and threats;*
8. Create *stronger* IAs;
9. Create/sustain *competitive advantage;* and
10. Undertake *corporate restructuring.*

Strategic Purpose for IP Valuation

Recently, a flurry of initial public offerings, powerful acquisitions and mergers, and high-profile infringement litigations have propelled IP into a vital position in global economics.[6] Unfortunately, however, several companies still fail to appreciate the *value of* and *risks to* their IP, even when the IP *accounts for a dominant share of the company's value.* Operating under growing market competition, limited resources, short product lifecycles, and under increasing pressures by the stakeholders for higher returns on investments (ROIs), companies must learn how to rapidly innovate, efficiently protect, and effectively leverage their IP to create "wealth" and "value." Inaction or indifference are not options in this market climate.

Thus, knowledge revolution and globalization have led to the creation and rapid growth of mammoth knowledge-intensive global companies (such as Amazon, Apple, Google, Facebook, Microsoft, and Infosys, etc.) whose *market value* is up to 85% *based on* ITAs and for whom IP is central to *competitive advantage* and *growth.* Consequently, today, *the new accounting standards* of the United States of America (U.S.A.) and other advanced nations *require firms to quantify and report the value of all identifiable ITAs in business transactions,* such as a merger or acquisition.[6]

Indeed, IP *valuation* and *communication of the value to stakeholders* have become just as important as *value creation, value protection, and value extraction measures for competitive advantage and growth* of a firm. Thus, Ben Bernanke, the former U.S. Federal Reserve Chairman, has recently applauded the *value of intangible capital*—noting that its accumulation accounted for more than half of the increase in U.S. output-per-hour during the past several decades.[7]

General Considerations for IP Valuation

It would be very helpful to revisit certain topics before taking up the *IP valuation approaches.* They are: *company's net worth/book value, market capitalization,* and *true market value of the firm.* These topics are discussed in great detail in Chapter 1. This discussion highlights the challenges involved in the *identification* and *accurate valuation of* ITAs.

Issues and Challenges in IP Valuation

It is important to point out the kinds of challenges one faces in the valuation of IP:[8]

1. First of all, the ITAs must be *identified*; indeed, some ITAs are very difficult to *identify* and even harder to *quantify*;
2. They must NOT be *in the public domain*. Hopefully, they are a part of TSs, CPK, or CIs. Of these, the CIs are most imprecise and hard to wrap hands around;
3. ITAs must be *useful* to generate *wealth* and/or *value*; and
4. ITAs must be of *transferable value* from one owner to another.

If an ITA fails to meet any of the above conditions, it would render itself useless. Consider some examples.

- A *research notebook* containing a *secret product formula* for a household care product may be considered as CPK, and hence an ITA. It will be useful to the company, only if the *secret product formula* is either *a patentable composition* and/or *useful to generate wealth and value, provided it can be guarded as a* TS. On the other hand, if the *secret product formula* is *very easy to copy* and/or *not patentable*, then the *research notebook* containing the formula will be worthless, and valuation of such IP will be futile; and
- If a company *licenses a patent for making a new product* that can join its existing line of products in the market, to increase revenues and market share—*it would be potentially valuable ITA for the company.* However, if the *licensed patent* requires "secret know-how" for making the new product, and the licensing company is unwilling to share that information, *then the licensed patent would be a wasteful ITA*, the valuation of which would be futile.

Defining the Deliverables for IP Valuation

Before selecting a particular approach for IP valuation, the company must define its goals:

1. What is the end-purpose of IP valuation? (corporate restructuring, corporate valuation for shareholders, strategic acquisition/merger/sale/licensing, etc.);
2. Who are the stakeholders of IP valuation?
3. What are the IP assets to be valued? (IPRs, TSs, CPK, or CIs); and
4. What is the degree of accuracy, transparency, and reliability being sought?

Thus, being clear about the deliverables is critical for choosing the IP valuation approach.

General Approaches and Methods for IP Valuation

Generally, companies use two approaches for IP valuation. These approaches are: *quantitative* and *qualitative* IP valuation approaches. While the *quantitative* IP valuation approaches focus on *calculating the economic value of* IP, the *qualitative* IP valuation approaches focus on *the analysis of IP characteristics* (such as the legal strength of the patent) and *uses of* IP. Figure 9.1 shows the details.

Quantitative IP Valuation Approaches

The "*quantitative* IP valuation approaches" comprise of the following methods[9]:

1. Cost-Based IP valuation methods;
2. Market-Based IP valuation methods;
3. Income-Based IP valuation methods; and
4. Option-Based IP valuation methods.

Each of the methods offer certain *advantages* and *disadvantages* for valuation of IP. It is important to recognize at the outset that there are *no perfect methods* for IP valuation that suit all possible scenarios.

Cost-Based IP Valuation Methods

These methods argue that *the economic value of an IP asset must be based on the costs incurred in its development*. Thus, "cost-based IP valuation methods"[9] assume *that the costs incurred in the production of the IP asset represent the value of the asset.*

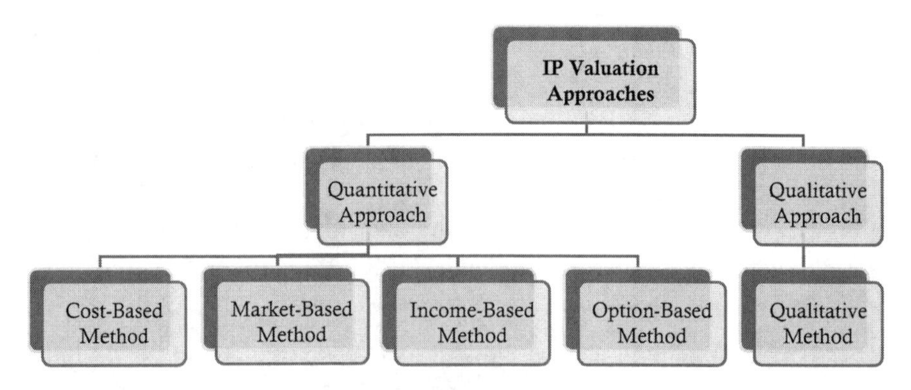

FIGURE 9.1
The methods for IP valuation.

Thus, there are *three* cost-based IP valuation methods:

- ***Historic cost method***: The "historic cost method" calculates *the costs sustained by the company in the development of IP, at the time the IP was developed*. Here, cost estimations are done by gathering all the costs associated with the purchase of IP;
- ***Replication cost method***: The "replication cost method" quantifies *all the costs* that would be needed to develop the same IP, in exactly the same way, at the present time, *including the costs incurred for unsuccessful prototypes*. In this method, today's costs are estimated to make a replica of the IP, including the labor and material costs involved for unsuccessful prototypes; and
- ***Replacement cost method***: The "replacement cost method" quantifies the costs that would be needed to develop the same IP, in exactly the same way, at the present time, *excluding the costs incurred for unsuccessful prototypes*. Thus, in this method, *replacement costs are estimated for an equivalent IP asset with similar use or function*.

Except the *historic cost method*, the cost-calculations employed by the other two methods are based on the "latest" cost-estimates, for example, not the "historical costs," but the "costs that would be incurred as of the valuation date."

These costs include:

1. *Direct costs*, such as on labor, materials, and management; and
2. *Opportunity costs*, relating to the *lost profits* due to *delays in market launch* or *lost investment opportunities* as a result of commitment to the IP asset.

Cost-based IP *valuation methods* are generally done by accountants complying with the Generally Accepted Accounting Principles (GAAP). This is how management researchers would refer to it. Common wisdom says that cost-based methods are *primarily useful for bookkeeping purposes* and sometimes *as a supplement to an income approach* (see below).

Market-Based IP Valuation Methods

These methods[9] argue *that the market place is a good source of information to decide the value of IP*. "Market-based IP valuation methods" *estimate the value of an IP asset by comparing recent transactions of comparable or similar* IP *assets between independent parties*. Thus, a company can estimate the value of an IP asset of interest, knowing the values of comparable IP assets bought/sold/licensed by independent parties. Accordingly, there are *three* market-based IP valuation methods.

- ***Comparable market value method***: In this method, the value of the IP is estimated by looking at comparable or similar IP transactions;

- ***Auction method***: In this method, the value of the IP is fixed at the highest bidding price; and
- ***Comparable royalty rate method***: In this method, the valuation of an IP asset is based on comparison of royalty rates when licensing similar IP.

These methods are used *when one wants to know the market value of* IP.

Income-Based IP Valuation Methods

Most people define the "value of an asset" in terms of its ability to generate ROI or, in other words, its ability to generate "future income." IP assets are no exception. "Income-based IP valuation methods"[9] measure the ROI of the subject IP in an effort to estimate *the value of the* IP. Income-based IP valuation methods differ based on the *valuation objective* and the *type of industry*, some of which are listed below:

- ***Discounted cash flow method***

  The "discounted cash flow (DCF) method"[9] is *the most widely employed of all the income-based valuation approaches*. In this method, the "net present value" (NPV)—of the future cash flows from the IP asset—is computed over the useful life of the asset *to estimate its value*. However, in the DCF method, two important factors must be taken into account: *the time value of money* and the *risk associated with the forecasted cash flows*;

  Experts who use this method take care of this issue in one of two ways:

  1. Through the use of a "special discount rate" set for the subject IP, which simultaneously accounts for both the above factors; or
  2. Alternatively, adjust the forecasted cash flows by taking into account *the riskiness of the asset* and *its varying riskiness over time*. When discounted at a *risk-free rate*, these account for *the time value of money*.

  Both these versions are widely used.

- ***Risk-adjusted net present value method***

  The "risk-adjusted net present value (ra-NPV) method" is a modified DCF method designed for valuation of the IP assets *specifically in the pharmaceutical and biotechnology industries*. To account for the risk, the ra-NPV method adjusts the cash flows at each R&D stage of the IP asset development *by assigning fixed probabilities based on established industry indicators*. For example, the ra-NPV method may decide to use *20% statistical probability for success* in the first R&D stage of clinical trials, it may use *10% statistical probability for success* in the second R&D stage, and so on. Thus, the ra-NPV method computes NPV and estimates the value of the IP asset, based on the cash flows carefully adjusted for risk *using experience-based probabilities for different stages*;

- ***Relief from royalty method***

  The relief from royalty method *assumes the value of the IP asset to be the rental charge that other companies would have to pay in order to use it.* Thus, the "relief from royalty method" estimates the "hypothetical royalty" that the company will have to pay *if it were to in-license the IP being valued, from a third-party.* The "hypothetical royalty" represents the "rental charge," which would be paid by the licensee to the licensor if such an arrangement were to be in place. Next, *a reliable sales forecast is required* to estimate the income that flows directly from the IP. Then, *the royalty rates are discounted at an appropriated discount rate;*

- ***Technology factor method***

  The "technology factor method" calculates "risk-free NPV" for the IP and multiplies this with a factor, called "technology risk factor." In the technology factor method, the technology risk factor is computed based on *the commercial strengths and weaknesses of the* IP. The goal of this method is to account for the *technical (in the case of technology), legal, market, and economic risks* of the IP being valued; and

Income-based IP valuation methods will be accurate and useful only if the following information is either available or accurately foreseeable:

1. *Income stream* from either product sales or IP asset licensing;
2. *Useful lifespan* of the IP asset;
3. *Specific risk factors* of the IP asset; and
4. *Discount rate* that is valid.

Option-Based IP Valuation Method

The "options-based IP valuation method"[9] assumes that a *patent has intrinsic value based on its projected cash flows discounted at the opportunity cost of capital.* This method takes into account *the intrinsic uncertainty in a business* and the *strategic management required for a successful patent-based business strategy.* The method estimates *the values of these factors* using the *Black-Scholes option-pricing model,* for which the inputs are as shown below:

Underlying asset value: the NPV of the IP's *future cash flows* over its life;

Exercise price: the NPV of the *fixed costs* that must be invested to commercialize the product and/or to maintain the patent's strength;

Time: the useful *patent life* left. **NOTE**: This method *does not consider* the future benefits beyond the patent expiration date;

Volatility: the *standard deviation of the growth rate* of the patent's cash flows;

Risk-free rate: the *risk-free* U.S. *Treasury rate* over the remaining life of the patent; and

Dividends: reduction of *the option's duration* due to competitive action, unforeseen delays, or other risk factors.

Qualitative IP Valuation Approach

The "*qualitative* IP valuation approach" provides a *value guide* for the IP asset on the basis of a qualitative rating and scoring of *value indicators* that either *positively and/or negatively impact the value of the* IP. Just as location, number of bed rooms, number of bath rooms, school district, etc. decide the *value of a house,* a combination of *impactful factors* (such as below) decide the *value of an* IP *asset.*

- **Patent information-related value indicators**

  In the case of patents, a *robust correlation* between *standardized indicators* in patent information and *patent value* is noticeable. Thus, the "value of a patent" can be assessed *based on the relative number of patents it evokes during the search and examination process* and *the number of citations a patent receives* indicating its innovation value. In other words, *the patent citations network* is a useful qualitative evaluation tool. Also, other factors—such as, *the number and quality of claims, the whole patent portfolio,* and the *competitive advantage* of the patent—may also be indicative of its value; and

- **Evaluation of value indicators**

  The *IP score software method* developed by the Danish Patent and Trademark Office for strategic valuation and management of patents—is an example of the *qualitative valuation method.* Thus, the *IP score software* enables companies to *internally estimate the value of its technologies, patents, and patent portfolios.* The *IP score assessment* of a patent (IP asset) consists of legal, technology, market, finance, and strategy *categories,* each of which is associated with 5–10 *index questions.* Each question relates to a different *value indicator.* Each question is rated 1–5 based on the *strengths and weaknesses of patents.* Together, the *value indicators* provide a holistic picture of the relative risks and opportunities of the patent. These can be processed into *different end formats*—tables and graphical forms—*for the purpose of strategic decision making.*

Advantages and Disadvantages of Various Approaches

The Advantages and Disadvantages of Cost-Based Methods

An important advantage of the cost-based methods is that IP *appears in the company's books* and this *creates/improves awareness of* IP *within the company.* Further, cost-based methods are particularly useful in the case of IP assets *whose benefits have not been established.*

There are also many disadvantages with the cost-based methods. To start with, cost-based methods operate *by totally ignoring the relationship between the cost of development and the future value of the assets.* This creates problems. In fact, there is ample evidence to show that IP *that incurs high costs* to produce *may not necessarily be highly valuable* to the business. On the contrary, there is also evidence to show that IP *that has been ignored and written down could suddenly pick up value.* Further, the calculation of historic costs is imprecise and unreliable in many circumstances. Thus, it may not be always possible *to precisely measure the costs incurred in terms of time, labor, and materials* in the IP development, as there may be real challenges in deciding what to include and exclude. In other words, IP assets whose value is determined by cost-based methods may not hold much promise for *return on investment.*

Advantages and Disadvantages of Market-Based Methods

The *comparable market value method* works well when *comparable exchanges of IP take place in the market place between independent parties and the transaction price information is readily accessible to the interested parties.* Unfortunately, however, the formal markets for IP are very limited and the actual market transactions are not usually accessible to public. As a result, the use of the *comparable market value methods* for valuation of IP as stand-alone methods, is limited.

Further, the use of the *auction method* is also limited. That is because ONLY *under ideal conditions,* there will be many potential buyers for the IP of interest, and all of them possess the necessary information about the IP to engage in bidding.

On the other hand, the use of *comparable royalty rates for IP valuation* is relatively more common, especially because strong IP holding companies and IP valuation firms are involved in collating *reliable databases* of *royalty rates in the industry* and *comparable transaction information.*

There are also many disadvantages to market-based approaches. Firstly, the uniqueness of IP makes direct comparisons between IP assets very arbitrary. These risks include—risks such as comparing the subject IP which one wants to immediately commercialize with another IP asset which has been traded, but has not been tested for commercial feasibility. In these cases, however, the subject IP could be undervalued. There are also potential distortions in royalty rate comparisons. Thus, royalty rates that have been set based on "return on R&D costs," "return on sales" or "industry averages" run the risk of *valuing costs* rather than *value*.

Advantages and Disadvantages of Income-Based Methods

The advantage of the *income-based methods* is that *the value of the* IP *is relatively easy to assess due to the availability of the market information and required inputs from the firm's financial statements*. Thus, a company may be able to identify and/or forecast cash flows with a reasonable degree of accuracy. However, in highly volatile environments, the methods may be difficult to implement, although they are conceptually robust.

Income-based methods always suffer from some degree of *uncertainty* and *subjectivity*. A major disadvantage of these methods is that a company must estimate both distant and uncertain cash flows, as well as the discount rate. For example, no information would generally be available to reliably predict the market potential and/or cash flows from untested IP assets. Furthermore, income-based methods *assume that all risks (such as legal risk, technological risk, environmental risk, etc.) can be lumped together and appropriately adjusted in the discount rate and the probabilities of success*, rather than dealing with each individually.

A significant drawback of the "relief from royalty method" is that a *hypothetical royalty rate can always be assumed, though it may never materialize*. Nevertheless, in specific circumstances this method is useful, especially if there are suitable comparable transactions involving third parties or industry standard royalty rates.

Advantages and Disadvantages of Qualitative Evaluation Methods

The main advantage of *non-patent value indicators* based on patent information is their relative *simplicity*. Once the relevant information has been researched and made available, *it is relatively easy to value the IP without the need for complex methods*. Another advantage is that the *data for the evaluation is often publicly available*. As a result, qualitative methods facilitate the comparison and ranking of IP within a company's own portfolio or against competitors' IP.

On the other hand, valuing IP using *patent information related value indicators* has many pitfalls. For example, a *mechanical count of citations* avoids in-depth analysis of their *real value*. Merely *focusing on simple counts* ignores additional information within the network of citations. Using *value indicators as a proxy for value* is *only as useful as the level of expertise* of those conducting the valuation. Finally, *the quality of information used in the IP score* dictates the quality and realism of the qualitative evaluation.

References

1. file:///C:/Users/usrac/Downloads/SSRN-id3009783.pdf
2. http://www.ipcentar.uns.ac.rs/pdf/IP.pdf
3. file:///D:/Book%20Projects/Taylor%20&%20Francis%20-%20IPRs%20for%20Scientists%20&%20Engineers/Manuscript%20in%20Preparation/Chapters/Chapter%209/References/Value%20Creation%20and%20Value%20Capture%20of%20IP.pdf
4. http://www.pro-tecdata.com/pdf/IAM-InfoIPProfit.pdf
5. (a) https://clarivate.com/wp-content/uploads/2017/10/IP_PatentPortfolioAudits_WhitePaper_A4_007.pdf; (b) https://www.fenwick.com/FenwickDocuments/IP_Audit.pdf; and (c) http://www.nishithdesai.com/fileadmin/user_upload/pdfs/Research%20Papers/Intellectual_Property__IP__Audit.pdf
6. https://www.marsh.com/us/insights/research/importance-of-intellectual-property.html
7. http://www.chegg.com/homework-help/questions-and-answers/chairman-ben-s-bernanke-conference-new-building-blocks-jobs-economic-growth-washington-dc--q3615640 (Retrieved on April 19, 2018)
8. http://www.wipo.int/export/sites/www/sme/en/documents/valuationdocs/inn_ddk_00_5xax.pdf
9. (a) http://www.ipcentar.uns.ac.rs/pdf/IP.pdf; (b) http://www.wipo.int/export/sites/www/sme/en/documents/valuationdocs/inn_ddk_00_5xax.pdf; and (c) https://faculty.darden.virginia.edu/chaplinskys/PEPortal/Documents/IP%20Valuation%20F-1401%20_watermark_.pdf

Index

Note: Page numbers in italic and bold refer to figures and tables, respectively.

Printed and bound by PG in the USA